Business management in fisheries and aquaculture

Ian Chaston, BSc., PhD., MBA

Fishing News Books Ltd
Farnham, Surrey, England

British Library CIP data
Chaston, Ian
Business management in fisheries and aquaculture
1. Fishery management
I. Title
639'.2 SH328

ISBN 0 85238 132 8

Published by
Fishing News Books Ltd
1 Long Garden Walk
Farnham, Surrey, England

Set and wordprocessed by
Medimedia Ltd, Godalming, Surrey
Printed in Great Britain by
Page Bros (Norwich) Ltd, Norwich

Contents

Preface

At the beginning of the seventies, fisheries and aquaculture were growth industries with new trawlers being launched, processing facilities expanded, large scale fish farming operations being planned and new areas of the oceans being exploited. Within ten years, this situation has changed. Trawlers lie idle, processing plants are closed and of the companies which still exist, many are in financial difficulties. This dramatic reversal of fortune has been attributed to the energy crisis, the depth of the world recession, political instability, the 200 mile limit and excessive fish stock exploitation. Yet with the exception of the latter two factors, other industries have overcome these problems, and even within the fishing industry there are companies who continue to prosper despite all such obstacles.

Explaining failure can be difficult, but determining the cause of success is quite simple. It occurs in those companies who have an effective management operation. Unfortunately in much of the fishing industry, the emphasis is on worrying about today and hoping that tomorrow will take care of itself. This is no longer a practical philosophy in the face of current world economic conditions. If the industry is to survive and grow, a more widespread adoption of improved business management practices is required.

The objective of this text is to present some of the important issues in modern management theory and to show how these can be applied to fisheries and aquaculture. The first half of the book examines the concept of strategic management. This involves managers in defining the strengths of the business, the environment in which the company operates, identifying a competitive advantage and developing plans to exploit this advantage more effectively in the future. The second section of the book covers selected issues within the functional specialisms of finance, marketing, production and personnel management. Financial issues include accounting principles, generation and analysis of accounting records, cost control and investment appraisal. The marketing chapters cover product positioning and the use of pricing, promotion and distribution to manage demand. The importance of planning and quality control is the focus of the production management chapters. Employees are an undervalued asset in the industry and therefore human resource planning and industrial relations are covered in the personnel chapters.

The text is designed for use by practising managers or public sector advisors in the industry, as well as to form the basis for a college level introductory syllabus on business management in fisheries and aquaculture.

1
The role of management

THE NEED FOR MANAGEMENT

On conference agendas and in the editorials of trade magazines, there are usually some references to the bleak economic future facing the fishing industry. The predictions of economic decline focus inadequate fish stocks, rising fuel costs and depressed market prices. Aquaculture, the newly emerging sector of the industry, appears to share these burdens with growth constrained by unforeseen technological obstacles, rising operating costs and, for some species, a production capacity in excess of market demand.

For those involved in fisheries and the aquaculture industry, causes of past failure are clear. Yet, with few exceptions, the industry cannot accept that the introduction of improved management practices focusing on better direction, co-ordination and control of company operations would result in improved commercial performance in the future. Suggestions that there is a need for a more effective management philosophy are usually met with the claim that change is not feasible, justified by statements such as 'fish catch is not predictable', 'the industry is composed of too many small independent operators', and in the case of aquaculture, 'the new technologies have to be perfected before systemised management can develop'.

The industry is a major economic component of society in many countries, and over the last three decades Governments have attempted to introduce various forms of assistance to help companies survive. The form of this public sector intervention includes quotas on foreign vessels to protect fish stocks, the introduction of new catching techniques, subsidies for boat owners and/or shore-based processors, and the construction of laboratories to study the biology of aquatic species. Acceptance of the validity of these solutions is ensured because prospective fisheries advisors are taught them at courses on industry management at educational establishments all over the world.

Keynesian economics hypothesises that Government assistance in the form of grants or subsidies can act as a catalyst for growth by supporting the commercial sector during periods of economic downturn. In a free market economy, however, this form of intervention can only exist for a limited time. If an industry is no longer capable of generating profit, then at the inevitable point when Government support is removed, bankruptcies will occur and member companies will cease to exist. Unfortunately for the fishing industry, the combined effects of inflation and world recession is now forcing Governments to withdraw or reduce their levels of financial support. In order to survive, therefore, it is essential that the industry begins to introduce more effective management philosophies and ceases to rely on public sector support mechanisms.

THE MANAGEMENT FUNCTION

One of the smallest commercial entities in the industry is the owner-operated in-shore trawler, and examination of the fishing operation can illustrate the various activities which constitute the function of management within a business. Prior to a new fishing season, the skipper will consider such issues as where to fish, on which species to concentrate, whether to contract landings to a single customer or sell the catch at port auction and how to finance early trips until revenue exceeds costs. All of these matters are related to the most important management function, that of 'planning'.

Before putting to sea, the skipper has to purchase fuel, re-equip the boat and possibly arrange for the delivery of ice. These actions comprise the management function of 'organising'. Except for the smallest of trawlers, the skipper will require assistance while out fishing, and therefore a crew will have to be hired. This is the 'staffing' function of management. Once on board these employees will need supervising, and thus involve the skipper in the 'directing' function.

During the planning phase, the skipper will have established some expectations for catch and revenue. As the season progresses he will review actual performance against these expectations and, if necessary, will introduce changes to his patterns of fishing. This monitoring of progress and revision of plans are part of the 'controlling' aspect of the management function.

In all likelihood the example skipper will have received no formal business education and if asked to comment on the management activities, would probably describe them as simply 'doing a job'. The terminology is unimportant. What matters is that the functions are being executed. The vital difference between a good and a poor skipper is that the latter has paid insufficient attention to such issues, or more formally is 'an inadequate manager'.

LEVELS OF MANAGEMENT

In the early life of a new fishery or an aquaculture venture, commercial development is usually undertaken by a number of small, independent operators. As output increases, this will attract larger companies who will introduce trawler fleets or grow-out systems, processing facilities and distribution systems. These new organisations can no longer be controlled by a single individual. The number of management staff will have to be increased, and this is usually accompanied by the appointment of individuals who will specialise in specific areas of the business.

The number and diversity of roles within a management group will be reflected by the establishment of a pyramid-type structure which is typically subdivided into five levels. The variation in staff responsibilities within the pyramid can be illustrated by a company involved in the processing of ground-fish for resale to wholesalers. At the base of the structure are the foremen in charge of the processing line who are responsible for organising the fish filleters. Above the foremen, who are known as 'first-line supervisors', will be senior foremen or shift managers. As 'lower management' in addition to sharing the organising function with the first-line supervisors, they are responsible for decisions on line staffing and directing the filleting operation. In turn, the lower management will report to 'middle management', whose main tasks are control and short-term planning. As even the smallest company will be involved in other activities such as accounting, procurement and product distribution, middle managers will also be responsible for directing other staff in the execution of these tasks. At the apex of the pyramid are 'top management' who are usually company directors. Their title is a misnomer because the key task of directors is not directing, but medium and long-term planning to meet the company objective of optimising future corporate performance.

MANAGEMENT TASK SPECIALISATION

The economic purpose of a firm is to engage in transforming raw materials into finished goods which can be sold at a profit. This is a continuous process. Even in the case of the above small company there are a number of tasks associated with the successful generation of profits and these can be classified into four major types:-

'Production', which involves the procurement and processing of the fish.

'Marketing', the location of a purchaser of the product, negotiating the sale and supervising delivery.

'Finance', which includes recording the flow of money during the purchase of raw materials, consummating a sale, and meeting the costs associated with the operation of the company.

'Personnel', the managing of the company's labour force.

As a company grows in size, each area of management will require the full time attention of one or more persons. This division of tasks is usually achieved through the creation of departments which specialise in one of the management tasks. Further increases in company size will lead to additional subdivision of tasks within each department (eg the subdivision of a marketing department into sales, product management and market research).

Individuals within each department will have received training and gained experience specific to their area of specialisation. This will permit the most effective execution of the department's tasks. A potential deficiency of the system is that as the level of specialisation increases, employees and managers will have less and less understanding of the roles of other departments. It is therefore vital that effective channels of communication are created which allow departmental heads to co-ordinate activities across all areas of the company so that everybody is working towards the same corporate goals.

The size and importance of a department will vary in direct relation to company need. In a company which operates a large trawler fleet and sells all landings at port auctions, the largest department will be boat operations and the smallest, marketing. Within this latter group, there will possibly be one or two sales staff who attend the auctions under the direction of a general sales manager. In contrast, in a company involved in acting as a selling agent or 'broker' for fish processors, the largest department will be marketing; and as the company acts in an agency capacity, there will be few production responsibilities other than an individual who communicates with the processors over the availability of products to be marketed.

MANAGING THE FUTURE

Over the centuries, the fishing industry has learned to accept that natural phenomena such as prolonged bad weather, a change in the direction of current flow or variations in water temperature can all have a significant impact on landings. Man in general, continually endeavours to gain greater control over his destiny, and fishermen are no exception. Improved gear technologies have been developed to stabilise declining catch, and distant water trawlers have been introduced when more local waters yielded insufficient tonnage of fish. Shore-based processing methods such as canning or freezing have been adopted to convert a perishable commodity into a storable food product. The influence of these initiatives are reflected in the catch statistics for the first half of the twentieth century. Given an exponential growth in world population over the same period, it is understandable that the industry has predicted for themselves continuing financial success as a key supply source of protein, while ignoring the warnings that many of the new methods would eventually lead to the decimation of fish stocks.

In the last thirty years, Governments have been forced to try and protect declining stocks through the introduction of conservation measures. But regretfully, these have often been too limited and too late. The industry will now have to learn to live with their inheritance of drastically reduced fish stocks in the foreseeable future, and will have to evolve new management philosophies to respond to shortages of raw materials coupled with the expectation of further public sector intervention to reduce total fishing effort.

In the past, commercial success was directly related to the size of the investment in trawlers and processing facilities. Even now, management's usual reaction to declining catch is to attempt to locate another area of the world to which their existing capital assets can be redeployed. If the industry is to survive, however, this approach has to be replaced by one which instead examines ways in which companies can be restructured to profitably exploit the fishing grounds which remain. The nature of these new corporate structures have yet to be precisely defined, although one successful approach is the replacement of large company-owned trawlers with smaller owner-operated boats. This may not be welcomed by fleet operators who perceive their loss of control of the fishing effort in an area, but if these companies are to survive they must accept the need for change instead of attempting to place obstacles in the path of re-organisation.

Aquaculture offers the potential to supplement raw material availability. Unfortunately, the growth in world population which sustained incremental sales for the traditional sectors of the industry has led to coastal areas suitable for low intensity culture systems being used as sites for new cities and industrial complexes. Furthermore the industrialisation of society to improve per capita income has led to the irrevocable pollution of the water supplies needed to operate aquaculture facilities.

Another area of world resource constraint, which must be recognised in managing the future, is energy. Current trawler designs are based upon experiences gained in earlier decades of low fuel costs. In the Gulf of Mexico, for example, it is estimated that a gallon of diesel fuel is needed to catch one pound of shrimp. With fuel costing $1.50 per gallon, it is simple economics to predict a minimum landed price for shrimp of at least $2.00 per pound. Trawler owners are only now beginning to deal with the energy issue, and it may be more than a flight of fantasy to speculate on whether we may see the return of sail or coal-powered drifters similar to those which formed the basis of the large European fleets before the introduction of diesel engines. Energy costs, however, are not just a problem for the fisherman; further processing and product distribution are both major consumers of energy, but there is minimal evidence to show that any attempts at conservation are being made within these sectors of the industry.

As well as supporting measures to conserve fish stocks, Governments also influence the industry through their national economic and social policies. During the sixties and seventies, the world economy permitted most Governments to experiment with expanding the welfare state by the re-investment of tax revenues and heavy public sector borrowing. Concern is now growing that these programmes can accelerate inflation, create a public debt of unmanageable proportions and, in countries most deeply affected by the current world recession, possibly bring them to the verge of bankruptcy. These facts are causing politicians to re-evaluate their national economic policies and this in turn means managements in the commercial sector now have to forecast the impact on their businesses of this change in Government philosophy.

An added dimension to the relationship between Government and industry is created by the increasing interdependence between the economic activities of all nations. Gone forever is a fishing industry which produces, processes and markets output all within the borders of a single country. Fishing is now dominated by multinational corporations, and management in even the smallest of companies will have to become more proficient in understanding the influence of global economic trends on future business performance. A processor of cod blocks in Nova Scotia who, only a few years ago, had to concern himself with competition from other Canadian processors, may now find that a change in world currency values can result in the American market being flooded with EEC output attracted by a short term strengthening of the US dollar. Given that even large banks have problems predicting variations on world currency markets, there seems little likelihood that a small fish company in Nova Scotia will be able to forecast such events accurately.

Changing social values must also be expected to influence the management task in the future. In the past the fishing industry was often the only source of employment in many areas of the world and this situation supported a very authoritarian style of personnel management. As the level of education has improved, individuals within the work force have become more concerned with their social rights and will continue to demand greater control of their own destiny within commercial organisations. In turn, this will necessitate managers developing a greater equality in their relationship with the work force, and being willing to accept greater employee participation in the management decision process. Unfortunately too many individuals in the industry are unable to accept this trend and speak with pride of the ability to resist unionisation in their trawler fleets and processing plants. Under these circumstances, change when it occurs is often under the banner of total social reform which can result in a drastic revision or even a collapse in management structure as can be seen in Central and South America.

The longer any industry has been established, the more resistant are management to accepting the potential impact of change on future performance. A tragic recent example is the Swiss watch industry who initially perceived the digital watch as a short-lived fashion trend and ignored the potential of incorporating electronic technology into their products. The same mistake was not made by the Japanese, who now dominate world watch markets through the introduction of products which utilise the benefits of the microchip to the full.

The fishing industry is not immune to the effects of technological change, and if companies are to survive it is essential to establish mechanisms to identify the potential impact of new technology on future performance. There is, for example, little evidence that fleet owners in the shrimp fishery are modifying boat construction plans, despite firm indications that aquaculture will be a more cost effective supply source by the late eighties. Also in the frozen fish sector, companies planning investment in new processing facilities appear to have paid insufficient attention to the impact of retort pack technology which is predicted to replace a major proportion of frozen food sales by the end of the century.

STRATEGIC MANAGEMENT

The prevalent style of management in the fishing industry is that of focusing attention on near-term issues, recognising change when it results in a crisis, and resolving major problems by drastic revisions in organisation structure or product range. In defence of this style, members of the industry would argue that many of the environmental variables which influence performance are unpredictable and therefore a more dynamic approach is not feasible. This defence is not unique to fisheries, it has been voiced in other sectors of the food industry. The interesting fact is that although the financial performance in most of the food industry exhibits violent fluctuations in profitability, there exists a select group of companies whose earnings remain relatively stable from year to year. Study of these particular operations will reveal that they have introduced a new approach to planning and control known as Strategic Management. The major elements of this process are as follows:

Stage 1: Establish the position of the organisation in relation to the business system of which it is a part. Identify the key components within the business system which may influence future performance.

Stage 2: Accurately define the current strengths and weaknesses of the organisation.

Stage 3: Specify the long-term performance targets for the organisation.

Stage 4: Formulate the business strategies to be utilised in the achievement of the performance targets.

Stage 5: Develop the business plans necessary to deliver these targets.

Stage 6: Execute the business strategies and plans which have been agreed.

Stage 7: Create a control system which permits an evaluation of actual versus planned performance. This system should also permit the organisation to identify the cause of performance shortfalls, and where necessary, suggest revisions in strategies or plans to attain business targets.

Within the fishing industry, there has been little research on the impact of strategic management on company performance. Over the last twenty years, however, studies in other industries have provided irrefutable evidence that strategically managed companies outperform organisations using more traditional management methods. In addition these studies indicate that strategic management is the most effective approach to achieving an improvement in the key performance areas of market share, profitability and return on investment.

2
Business systems and company evaluation

DEFINING THE BUSINESS SYSTEM

The objective of all commercial organisations is to convert procured raw material into a more valuable product which can then be sold at a profit. If this activity is to be consistently executed, it is necessary to identify the key variables which could influence the organisation's performance. Defining these variables is the first step towards building the business system model.

The modelling process can be illustrated by studying a small shrimp processor located in Texas on the Gulf of Mexico. During the period of June to November, the company buys fresh, shell shrimp tails (known as 'green headless shrimp') from local, independently-owned trawlers. During the rest of the year, raw material is purchased in the form of bulk pack frozen shrimp from Mexico and Central America. Upon arrival at the plant, the green headless shrimp is graded, washed and repacked into 5 lb boxes to be traded in the commodity market or partially peeled. The partially peeled shrimp is used to make breaded shrimp for sale under the company brand name or packed for one of three food service wholesaler customers under their brand name.

The key variables influencing the company's activities are:

(1) Green headless shrimp, which is the raw material for all of the company's products. Availability is mainly dependent upon landings from the Gulf of Mexico fishery. Landed prices will reflect the level of catch and will in turn directly control the company's finished goods prices.

(2) The labour force, and as shrimp processing is labour intensive, this represents the most important internal cost variable.

(3) The short-term financing needed to fund raw material procurement and the holding of finished goods. The fishing industry faces seasonal landings but year round sales: this imbalance between the timing of supply and demand is usually solved through overdraft borrowing from a bank.

(4) Market demand for the company output, which will define the annual sales revenue. The demand variable is composed of two subcomponents; market acceptance (or brand loyalty) for the company's product and overall market demand for all of the shrimp industry's output. These subvariables have differing effects on the company product line. Overall market need is the dominant factor influencing sales of the 5 lb green headless packs traded as commodity goods. The important subvariable for breaded shrimp sold under the company brand name is brand loyalty.

(5) Competition from other shrimp processors in the immediate vicinity who are alternative employers for the available labour force and will also bid against the company on fresh shrimp landings. In conjunction with these local processors, competitors with plants elsewhere in the USA will all be attempting to out-market the company to sell their products to supermarkets, wholesalers and restaurants.

These five variables will have the most direct influence on company performance. The management must therefore fully understand the nature of their effect and ensure effective monitoring in order to identify any changes before they become an obstacle to continued commercial success. With the aid of the computer and operations research methods, some industries have been able to develop equations which predict the impact of changes in key system variables. This concept has not gained widespread acceptance in the fishing industry, although even in the example of the shrimp processor, it is possible to hypothesise on the probable form of such equations as follows:

Future catch (C_{t+1}) can be approximated using a simple time series of

$$(C_{t+1}) = a \sum_{i=1}^{n} C_{t_n} \div n$$

where

C_t = total catch in year t
n = the number of years in the time series
a = a calculated constant

Future price is related to expected supply and the relationship between price and supply can be obtained from a regression analysis of previous years. This will yield the equation

$$P_t = b + cC_t$$

where

P_t = average price in year t

b and c = calculated regression coefficients

Hence having estimated the expected future catch C_{t+1} the expected price P_{t+1} can be calculated because

$$P_{t+1} = b + cC_{t+1}$$

Labour supply usually only changes over a long period and hence a predictive equation for this variable would have little application over the normal planning time profile of three to five years.

The market demand variables require an analysis of total industry sales and the company sales performance will also be influenced by promotional expenditure relative to competition. Total industry shrimp sales, both green headless and breaded, can be expressed as a function of total industry fish and seafood sales (TFS) by regression equations of the type

$$TGH = d + e\ TFS$$

$$TBS = f + g\ TFS$$

where

TGH = total industry green headless sales

TBS = total industry breaded shrimp sales

d, e, f, and g are regression coefficients

Company sales can be estimated from the equations

CGH = (promotional expenditure ÷ industry expenditure) x competitive coefficient

where CGH is the company green headless sales, and company breaded shrimp sales (CBS)

CBS = (promotional coefficient ÷ industry expenditure) x competitive coefficient

The competitive coefficient in both equations reflects the implicit strengths of the company relative to competition. A coefficient $\geqslant 1$ would indicate a strong position in the market, whereas a value $\leqslant 1$ will exist if competitors are more effective at marketing their products.

Systems theory states that no system, except the universe, can be considered an independent, closed entity because it is also a subsystem of a larger system. If each system is combined into larger systems, the complexities of these structures would soon approach the point where definitions of the numerous interacting variables would become impossible. To overcome this problem, the usual approach in business system models is to specify that the company system is also a subsystem of a larger structure known as the macroenvironment in which the key variables are the economy, technology, public policy and social culture. Their specific influence is more difficult to measure than intra-industry variables of the type described in the shrimp processor example. Nevertheless the potential impact of changes in the macroenvironment on a company is very significant. Hence as part of the modelling process it is essential to establish the importance of variables external to the industry business system which could influence performance.

The difficulties of forecasting events in the macroenvironment is demonstrated by the failure of even Governments to identify the recession at the end of the seventies until well after its onset. The impact of this adverse economic trend on the fishing industry are now all too apparent. During the twenty years prior to 1980, the industry had enjoyed economic growth, and despite production costs rising faster than the average rate of inflation, the market was willing to accept continued increases in price for fish and seafoods. This led the industry into further investment in catching, aquaculture and processing facilities. The new economic scenario is of no further increase in per capita income, rising unemployment and instabilities amongst most of the world currencies. To survive, the fishing industry will have to comprehend the implications of these economic conditions which may prevail for at least the balance of the decade.

Public policy is a variable which reflects the influence of Governments on society. The form of the influence can be direct through legislation on an issue such as work safety in an industry, or indirect, from actions related to a Government attitude on the need to change overall social conditions. At the beginning of this century, Governments were persuaded by the electorate in many countries that legislation was needed to improve working conditions. This was later expanded into safety for the consumer. After World War II, there was further pressure on many Governments to legislate on job security and terms of employment. The costs of such programmes were shared by the employer, who in turn reflected them in prices, and by Governments who funded through taxation the bureaucracies required to enforce their legislation. World economic growth permitted such policies, but some Governments are now facing the problems of supporting these programmes as the recession places an excessive financial burden on both industry and the public sector. This is therefore another area of the macroenvironment which could undergo major change in the next few years.

Possibly the most ignored environmental threat is that of technological change. Managers in most industries believe that technical innovation is most likely to occur from within the framework of the companies which comprise their own industry. Unfortunately history has repeatedly shown that there are major risks in this assumption. Fish curers ignored the invention of cans, canners in turn the advent of plate freezers. Further at the current time, the industry is apparently ignoring the advances in aquaculture being achieved by 'non-fish' multinational companies in the solution of what previously were considered insurmountable technical problems associated with the culture of certain aquatic species. These are but three examples of technological change originating outside the industry, and management must therefore place greater emphasis on technological forecasting if errors are to be avoided in the future.

Social culture is an extremely diffuse variable which is often impossible to measure until the change is almost complete. The outcome is demonstrated by the change in market demand for certain products. In the USA, for example, many of the ethnic demands for cured fish have disappeared as first generation immigrants from Europe are no longer a major element in the population. Another area of influence of social change is in employee behaviour. When an economy is underdeveloped, the tendency is for workers to accept the 'work ethic' and not question authoritarian supervision, poor working conditions and low wages. As the overall standard of living improves, gradually the majority of society will favour organised wage negotiations, and employees will expect to participate in management decision processes.

THE COMPANY APPRAISAL PROCESS

Given the need to generate profits in order to survive, the logical first step in an evaluation of a company is to examine trends in financial performance. Summary data are provided by the year end accounts. The profit and loss statement describes sales, cost of goods sold and net profit after operating expense. In many countries it is an accepted practice that the accounts also contain the prior year figures. Although this history is useful for comparison purposes, in order to complete a totally meaningful financial appraisal it will usually require a review over a much longer period of somewhere between three to five years.

In the creation of any business, it is an objective that the funds invested will generate a level of profits in excess of the alternative proposition of placing funds in a lower risk asset such as Government securities. Hence another component of the appraisal process is to measure the company's Return On Investment (ROI) using the formula:

$$\text{ROI} = (\text{gross profit} \div \text{sales}) \times (\text{sales} \div \text{company assets} - \text{current liabilities})$$
$$= \text{gross profit} \div \text{net assets}$$

It should be recognised that annual accounts are a summary of performance, and for companies engaged in more than one type of activity an analysis by business type may be more informative than an assessment restricted to overall corporate performance. This can be demonstrated by a review of a fish processing company which markets branded frozen, unprocessed fillets and breaded fish portions to supermarket chains and also similar products to the food service sector (eg hotels, restaurants) through three of their own food service wholesale distribution centres located in large cities.

From Table 2:1 it would appear that over a five year period performance has remained unchanged, with a small increase in total sales resulting in a net profit improvement of £1 700. Many companies in the current economic climate would be satisfied with this level of business stability and if no further analysis was carried out, it would not be surprising for management to make no revisions to their plans.

In contrast, an examination of the same data by area of business activity (Table 2.2) reveals a very different picture. Sales volume for fillets and breaded portions in the

Table 2.1 A comparison of annual financial performance for a fish processor

	1980 £	1984 £
Sales	380 000	400 000
Cost of goods	311 600	330 000
Gross profit	68 400	70 000
Operating expense	36 850	36 750
Net profit	31 550	33 250

retail sector have fallen by 14.0% and gross profit margin for breaded fish is down by 5.0 percentage points. Some compensation is provided by a reduction in operating expense, but nevertheless the retail activities are showing a combined decline in net profit of £4 200. Fortunately the food service wholesale operation has achieved significant success over the same period and the increase in net profit of £5 900 has more than compensated for the downturn in the retail area. On the basis of these results, a more effective management decision would be to direct a greater proportion of future resources towards the expansion of the food service business. This is a very different decision to that reached following an analysis restricted to overall performance.

Accounting records will only identify problems; they will not explain the causes. Hence the second phase of an appraisal is to isolate those factors which have caused the trends in financial performance. It is then possible to identify the impact of environmental variables on the company and to pinpoint the strengths and/or weaknesses within the various departments which comprise the organisation.

Company performance is heavily influenced by two factors in the business environment: the trend in total market sales and the company's ability to execute the marketing function relative to the same activity by competitors. The above processor example is based upon a Canadian corporation, and over the period 1980 to 1984 the North American fish and seafood market has exhibited significant growth in total sales. It must therefore be concluded that the company's marketing activities in the retail sector have been poorly executed if sales declined in the face of overall market growth. This contrasts with the wholesale distribution operation where sales have increased at a rate in excess of

Table 2.2 A comparison of financial performance by business area for a fish processor

	Retail business sector			
	Fish fillets		*Breaded portions*	
	1980	*1984*	*1980*	*1984*
Sales	230 000	200 000	90 000	75 000
Cost of goods	196 100	170 000	67 500	60 000
Gross profit	33 900	30 000	22 500	15 000
Operating expense	18 850	16 000	12 600	8 250
Net profit	15 050	14 000	9 900	6 750

	Wholesale distribution operation	
	1980	*1984*
Sales	60 000	125 000
Cost of goods	48 000	100 000
Gross profit	12 000	25 000
Operating expense	5 400	12 500
Net profit	6 600	12 500

market trends, thereby indicating a management ability superior to that of their competitors in this market sector.

An internal review of corporate capability is usually carried out by studying each of the functional areas of management, ie marketing, finance, production and personnel. Marketing is the function which interfaces with the environment, and some conclusions on the execution of this task will have been reached during the earlier study of business performance relative to competition. A fuller understanding of the strengths and weaknesses, however, can only be obtained from examining such issues as:

- How do the current range of company products meet market need and/or duplicate competitive offerings?
- How effective is the pricing strategy in supporting the chosen value/quality positioning in the market place?
- What plans exist for the introduction of new products?
- Do the various components of the promotional programme represent the most cost effective system to achieve the company sales objective?
- Have the sales force established an adequate presence for company products within the chosen distribution channels?

- Is sufficient attention given to the use of market research to generate information that can assist in the problem resolution process in the future?

The production function can be represented by very different activities depending on the company concerned and its position in the business system. A generalisation is that the function covers the responsibilities of acquiring raw materials and modifying them to a form suitable for resale. In the case of a trawler, this could include catching, gutting, icing and boxing. For aquaculture the raising and harvesting of a crop is the equivalent process. An assessment of effectiveness will cover such issues as:

- Is the potential production capacity able to support future demand for output?
- How efficient is the company at minimising production costs compared to other members of the industry?
- Is sufficient attention given to the introduction of new technology to keep costs at a minimum in the future?
- Do adequate mechanisms exist to ensure the most favourable decision being made in the key areas of raw material procurement, production scheduling and inventory management?

The personnel and finance functions are concerned with the application of two other input resources, people and money, which together form the core of the organisation. Therefore an evaluation of these functions will centre upon (a) whether the existing human and financial assets are being optimally utilised and (b) ensuring that future growth is not constrained by an inability to acquire additional supplies of either of the asset types.

Given the rapidly changing nature of the business environment, the appraisal should cover the company's capability for research and development. The traditional sectors of the fishing industry have tended to ignore this activity and have relied upon public sector institutions as a source of new ideas on catching methods and equipment. The more technical nature of the aquaculture process has required much larger budgets initially being allocated to research. However, companies in this industry sector will usually de-emphasise research and development upon identification of a suitable culture technique. In order for the fishing industry to combat the competitive pressures from other sectors of the food industry, a much greater proportion of profits should be reinvested in research in the future.

Having isolated the company's strengths and weaknesses, it should be possible to identify which of the internal resources have the capability to support a position of superiority over competition in one or more market sectors. Once this 'competitive advantage' is defined, it provides the foundation stone for the company's business strategy. In the above fish processor example, the management identified an advantage in the food service sector. However, further analysis would be necessary before a conclusion is reached on why this area of operations has been successful. Review of the market would show that most wholesalers are not closely affiliated to a primary processor and have to acquire their fish supplies from a variety of sources. This creates uncertainty in their procurement programme which will impair their relationship with customers. The example company, through vertical integration of the processor and wholesaler activities, has unwittingly created a competitive advantage which can be the basis for continued growth in the future.

3
Objectives and strategies

CORPORATE OBJECTIVES

It would be unthinkable for a pilot to depart without knowing in which direction to fly. In the same way it is difficult to operate a company effectively without having a performance target for which to aim. This target or 'objective' is usually stated in terms of sales, profit and ROI.

The assessment of competitive advantage and environmental trends will have provided insight on the company's probable future. If the outlook is bleak, the company is likely to establish an objective of survival, probably emphasising cost reduction and, where possible, reduction in fixed assets in order to sustain the current ROI. A very similar objective might be established by a company whose market appears to be mature and whose share of industry sales unlikely to change. This is often known as a 'satisfying objective'.

The danger of these objectives is that management tends to concentrate on cost control and sustaining a 'status quo' leaving the company vulnerable to a competitor who may introduce an aggressive sales expansion policy. It is preferable therefore, to establish a growth orientated objective relative to current performance, but it is important that the envisaged magnitude of growth is realistic. There is little point in defining an objective which is impossible to attain, for in this case, lower management will lack the confidence to develop the type of plans necessary to achieve the required financial goals. This philosophy can be illustrated using the processor example in chapter 2. Overall performance is trending downwards and with the current ROI at 20%, lower management would consider an objective of 25% for all areas of the operation to be a non-feasible target against which to plan. However, if the objective was restated as achieving a 30% ROI for the wholesale distribution operation and sustaining the existing ROI of 12% for the retail business, on previous trends these would appear to be reasonable targets for the future.

The more forward thinking companies recognise that they have certain responsibilities to employees and the community in which the business is located. In the last twenty years, therefore, these issues have been included in their corporate objectives. The employee objective will encompass earnings (eg profit-related bonuses for directors and productivity incentives for staff), opportunity for internal promotion and improvements in the conditions of employment. Social responsibility objectives cover such issues as the company's desired public image, contributing to an improvement in overall economic conditions in the area and minimisation of pollution from the production processes.

SURVIVAL STRATEGIES

Having settled for a financial objective of survival, the most probable strategy is to concentrate on some issue related to cost reduction. This type of strategy is frequently deemed appropriate by trawler fleet owners when catch has declined for either biological reasons or as a result of quota restrictions introduced by a Government. Strategies of cost reduction can include moving from a crew wage to a crew share of catch earnings system, or tying up the boats except during peak landing periods. These are short-term solutions. A more effective approach is to reduce indirect costs by improving the efficiency of the catching operation. The introduction of new gear technology is one possible strategy, as demonstrated by the Canadians in their move from bottom to mid-water trawling in the seventies and, later in the same decade, the return to beam trawling by boats in the southwest of England.

Where the financial survival objective is defined in terms of ROI, this is usually only attainable by reducing the fixed assets through the sale of capital equipment such as the trawlers or shore-based facilities. Tragically this has been considered the only solution in parts of Europe over the last five years, leading to the total cessation of activity in certain places (eg the once thriving deep sea ports of Hull and Grimsby in northeast England).

GROWTH STRATEGIES

A company's greatest expertise will lie within the sector of the industry in which it currently operates. Therefore it is in this area of the business system that all planning exercises should commence. As shown by Figure 3.1, there are four possible areas of opportunity:

(1) Expanding sales for current products in the current market sector.
(2) Expanding into a related business area with existing products.
(3) Introducing related new products into the existing market.
(4) Introducing related new products into a related business area.

A company should only consider these strategies if they are certain that the chosen market sector offers sufficient total sales and that the company products have a perceivable advantage over competition. An application of this type of planning matrix is provided for a crab company which operates a small trawler fleet and currently markets bulk pack, fresh

PRODUCT DIMENSION

MARKET DIMENSION		EXISTING	RELATED
	EXISTING	Current business activities	New product in existing market
	RELATED	Current product introduced into a new market	New product introduced into a new market

Fig 3.1 Existing and related business area planning matrix

crabmeat to hotels within a twenty mile radius of their processing plant (Figure 3.2).

		PRODUCT DIMENSION: EXISTING	PRODUCT DIMENSION: RELATED
MARKET DIMENSION	EXISTING	Bulk pack, fresh crabmeat sold to local hotels	Fresh crab plate sold to food service customers
	RELATED	8 oz retail packs of crabmeat sold to supermarkets	Stuffed crab products sold into the retail sector

Fig 3.2 Application of the existing/related business planning matrix

DIVERSIFICATION STRATEGIES

An alternative growth path is to diversify into areas outside the current areas of immediate business. There is potential for integrational, business related or conglomerate diversification.

Integrational Growth

Every company is part of a multi-level industry structure as shown in the earlier section on business systems. It may be possible for a company to move vertically into adjacent levels of the system (eg a processor opens wholesale distribution branches) or into new business areas within the company's present level by horizontal integration (eg a crab processor who also starts to pack scallops). These alternatives are described in Figure 3.3a and an example provided in

BUSINESS TYPE HORIZONTAL AXIS

BUSINESS TYPE VERTICAL AXIS	RELATED	SIMILAR	UNRELATED
RELATED			
SIMILAR			
UNRELATED			

Fig 3.3a Integrational growth planning matrix

BUSINESS TYPE HORIZONTAL AXIS

BUSINESS TYPE VERTICAL AXIS	RELATED	SIMILAR	UNRELATED
RELATED	Land other groundfish species	Land pelagic species	Raise marine species using aquaculture
SIMILAR	Further processing of landed groundfish	Further processing of pelagic species	Further processing of aquacul- ture output
UNRELATED	Open retail fresh fish stores	Create chain of food service wholesalers	Establish chain of fish and seafood restaurants

Fig 3.3b Example application of integrational planning matrix - a company currently operating a trawler fleet catching cod

LIMIT:
156 issues per customer.

YES! Renew my TIME at the low sale rate of:

☐ 3 years (156 issues) at 89¢ an issue.
I save 55%*.

I prefer:

☐ 2 years (104 issues) at 95¢ an issue.
I save 52%*.

☐ 1 year (52 issues) at 97¢ an issue.
I save 51%*.

☐ 6 months (26 issues) at 99¢ an issue.
I save 50%*.

NAME

ADDRESS/APT. NO.

CITY/STATE/ZIP

☐ Payment enclosed. ☐ Bill me in full.
☐ Bill me in 3 monthly installments.

*Off the $2.00 cover price.
TIME's basic rate is $1.12 per issue.

TDARKJ5

Figure 3.3b for a company currently operating a trawler fleet which concentrates on catching cod.

Conglomerate Growth

It is an understandable tendency of most managers to restrict their planning horizon to areas of business where they have some experience. This attitude, however, can limit organisations to an area of industry which has no potential for growth. A planning exercise should not be considered complete until a review has been undertaken of whether greater profit opportunity exists by expanding into a business area outside of the industry of which the company is a part.

The alternative growth path of conglomerate diversification may become increasingly popular in the fishing industry as traditional sectors recognise the downward profit trend in their current business activities. Two examples where this solution has already been introduced are frozen fish processors who have moved into other areas of the food business, and trawler fleet operators who have redirected resources into servicing rigs in the off-shore oil industry.

EVALUATING STRATEGIC ALTERNATIVES

Although a company may be capable of defining a large number of alternative opportunities, it is unlikely that sufficient management and/or financial resources would be available to pursue more than one new strategy at a time. It is therefore necessary to decide which growth path is the most likely to achieve the defined corporate financial objectives.

A strategic decision matrix of the type in Figure 3.4 permits the evaluator to assess each alternative in relation to two important variables - company resources and market opportunity. The optimum strategy is that which offers high market opportunity and can receive a high level of support from existing company resources. This may prove to be an excessively restrictive decision rule and therefore many companies will also include the related matrix cells, namely medium resources/high market potential, high resources/medium market potential and possibly even medium resources/medium market potential.

Any strategy which involves moving from the firm's existing area of business will increase the level of uncertainty about the accuracy of the forecast for future

MARKET OPPORTUNITY

COMPANY RESOURCES	LOW	MEDIUM	HIGH
LOW			
MEDIUM		Sub-optimum strategy Choice	Optimum strategy Choice
HIGH		Sub-optimum strategy Choice	Optimum strategy Choice

Fig 3.4 A strategic decision matrix

financial performance. To accommodate this factor, many analysts add a risk element to their performance target calculations in the form of a probability assessment of success or failure.

For example a shrimp farm considering whether to expand output by building more extensive culture ponds or by introducing new technology in the form of intensive raceway grow-out systems, might estimate annual revenue and outcome probabilities as:

EXTENSIVE		*INTENSIVE*	
Revenue('000)	*Probability of outcome*	*Revenue('000)*	*Probability of outcome*
+ 300	0.2	- 700	0.3
+ 600	0.6	+ 900	0.5
+1000	0.2	+1300	0.2

The expected revenue for the alternative approaches can be calculated from the sum of the revenue and probability data because

$$\text{Expected revenue from the extensive system} = (300)(0.2) + (600)(0.6) + (1000)(0.2) = 620$$

$$\text{Expected revenue from the intensive system} = (-700)(0.3) + (900)(0.5) + (1300)(0.2) = 500$$

If the farm operates under a decision rule of maximising expected income, then the chosen strategy will be to expand output using the current system of extensive ponds.

THE STRATEGIC DECISION

The responsibility for generating and evaluating strategic alternatives may be considered a prerogative of top management in some companies and delegated to lower management or a corporate planning group in others. Whichever is the case the final decision on which strategy to follow must lie with the chief executive. The nature of this individual's decision will be influenced by whether he is a conservative manager tending to select low risk propositions, or a risk taker who will choose a path offering maximum possible financial return despite possible indications of high risk.

The chief executive who has worked in the same organisation and industry sector will tend to favour a strategy based upon past success. Under these circumstances, when changing conditions force a need to introduce new solutions, the tendency is to graft on new substrategies to existing plans. Only when there has been a major change in business conditions will the leadership begin to accept a more radical solution. This effect is very apparent in the fishing industry where all too often, firms have to be almost bankrupt before major revisions in plans are introduced. A recent example is provided by the distant water trawler operations in Europe, which for many years refused to accept the implications of declining stocks and the probable introduction of restrictions in catch activity. Many firms continued to invest in new, often larger, freezer boats during the sixties and seventies even though there were strong indications of an uncertain future for these assets. When the 200 mile limit and more restrictive quotas were introduced, these companies were left owning some hulls which had only completed one or two seasons fishing.

The aquaculture sector is fortunate in that many of its senior managers have come from other industries and therefore have a greater ability to introduce innovative approaches to the development of new strategies. This is shown by groups willing to alter culture methods (eg switching from pen-raised to ranching of salmon), evaluate new species (eg South American herbivorous fish), and where business obstacles are encountered, the ability to develop effective solutions (eg salmon farmers who established a mail order operation to distribute smoked fish when faced by resistance in traditional distribution channels to carry their product).

Even in aquaculture, some organisations who have been established for a few years have tended to reduce their level of innovation. For example, the trout industry in some countries has reached the point where market demand is exceeded by supply. Instead of accepting this situation and seeking new revenue paths such as integrational growth or diversification, many firms in an attempt to increase per capita consumption, have responded by investing in inadequate promotional programmes, or even more disastrously, have embarked on unprofitable levels of price cutting.

Once 'strategic ossification' has become entrenched within an organisation, the ability to react to change is severely reduced. An extreme solution practised by some industries is to replace senior management on a frequent basis. This tends to weaken an organisation over the long-term, and hence the more forward-thinking companies have evolved systems in which managers are encouraged to continually review their strategic decision relative to environmental trends. Management are also trained in techniques of creative thinking and promotion is in part based upon an individual's capability to evolve innovative solutions to business problems. In a society where change is the only constant, this can be the only effective approach if a company is to survive in an increasingly complex world.

4
Departmental strategy formulation, planning and control

DEPARTMENTAL STRATEGIES

An organisation is usually composed of specialised subunits or departments with differing roles. In order for each department to fulfil its responsibilities, it will be necessary to develop detailed subsystem strategies which together result in the achievement of the overall corporate goals.

This concept can be illustrated by looking at the departmental strategies for a processor who is currently smoking and distributing premium quality mackerel fillets. A review of the external environment has revealed future problems due to the reduced availability of locally caught raw material and increased competition from lower-priced, imported smoked fish. The revised corporate strategy is to diversify into other smoked foods but, in order to sustain the current ROI growth rate of +5% per annum, not to permit any incremental investment in new resources.

Each department will assess this diversification decision in relation to their current activities and begin to plan a new strategy within their department to fulfil their new responsibilities. The marketing group will implement research on market demand for other smoked products to determine which items provide sales opportunities. Given the constraint of no incremental investment, the additional sales could not be generated by geographic expansion as this would entail hiring more sales staff. Opportunity evaluation will therefore have to be restricted to the existing customer base. In this example, the marketing group decided on a strategy of adding smoked salmon, prawns and scallops to the product line. Within the near future, they also hoped to introduce non-fish products such as poultry, meat and cheeses.

Decisions in one department will have ramifications in other areas of the company. The marketing strategy of product line expansion will require new production capability, and therefore during the departmental strategy development process it is essential that all groups are kept fully informed of one another's new proposals. In many instances, the primary source of influence in substrategy planning will be the marketing department because their decisions will control the direction of thinking by other departments. The product line expansion plan in the smoked fish processor example will cause the production group to develop a strategy in which existing equipment and employee expertise can be utilised to manufacture the new items. A move from mackerel to a variety of products will involve a change from continuous, single species processing to a batch system capable of handling a variety of raw materials with differing physical characteristics.

The personnel department is responsible for the organisation's human resources, and their substrategy will reflect the staffing needs associated with departmental plans. The constraint of no incremental investment in the smoked fish example would mean a personnel recruitment strategy of no additions to the workforce except on the basis of replacements for departing staff. Under these circumstances, the focus of the personnel strategy will be on training programmes to ensure that current employees are able to handle the increased complexity of tasks associated with the product line expansion.

The provision of adequate financial resources to support the acquisition of raw materials, fund capital investment programmes and optimise revenue inflows against expenditure outflows is the responsibility of the accounting department. Achievement of these objectives will be reflected in their substrategy. In the smoked fish example, the lack of new investment will mean that no additional borrowing will be permitted to fund the introduction of new products. Consequently the accounting department strategy will have to focus on mechanisms to use internally generated sources of funds to support the marketing department's plans.

During the corporate strategy planning stage, the aim is to develop an overall company concept, and it is assumed that departmental strategies will provide additional evidence to support the validity of the overall plan. Occasionally the departmental activity may reveal inconsistencies of logic. Where this occurs and identified obstacles prove insurmountable (eg a diversification programme which cannot be executed without incremental investment), it may be necessary to identify an alternative, more feasible, strategy.

DEPARTMENTAL PLANNING

Given the complex interactions between departmental strategies, their execution can only be achieved if a detailed brief of required actions is prepared. Each department will firstly have to decide the resource demands for manpower, equipment and materials. It is rare that incremental financial support is approved by management and therefore it is usually necessary to assess the implications of transferring resources from other areas of activity. A secondary benefit which may accrue from this exercise, is that financially marginal projects are identified where immediate termination will improve company profitability.

The second phase of the planning process will deal with the specific sequence of events which will lead to departmental strategy execution. This can be illustrated by the product introduction proposed by the marketing group in the smoked fish company example:-

(1) Product concept finalisation - From the preliminary ideas developed during the earlier work on strategy definition, final concepts can be prepared and product prototypes designed.
(2) Market research - It is necessary to quantify market response to the prototypes using placement studies and user surveys.
(3) Product specifications - The research results will provide the data for deciding the final product specification, manufacturing formulation, pricing and packaging.
(4) Promotional plan - To achieve rapid market acceptance promotional activity will have to be planned which can achieve user awareness and trial purchase.
(5) Sales training - The product launch will rely heavily on sales staff who require training on (a) product usage, to respond to customer questions, and (b) how to execute those elements of the promotional programme for which they will be responsible (eg trade shows, product demonstrations and introductory allowances).
(6) Product launch - This represents the cumulation of the earlier events defined above.

(7) Follow-up programmes - Long-term success will depend upon converting initial trial by customers into a permanent product adoption. To achieve this objective will require promotional events to build repeat purchase and further research to monitor customer response after the initial product trial period.

The third phase of planning is to prepare a timetable for the required sequence of events to identify which stages are dependent upon the input from other departments. The tendency for most groups is to underestimate the time that others will need in order to provide their input. Where departments cannot reach agreement on an acceptable schedule, senior management may be called upon to arbitrate in order to resolve the issue. In the smoked fish example, the marketing group will need to estimate, in conjunction with the production group, how long the development of prototypes will take and, at stage three, jointly agree the time to be allowed to finalise the manufacturing specifications for the new products.

The final stage of the planning is to ascertain whether the existing organisational structure is capable of executing the new strategies. Where there is little expected change in business activity, few revisions will be required. Until this point, lower management and the majority of the workforce will not have been involved in the planning process, and a staff briefing programme should be scheduled, accompanied by the preparation of written procedure documents for use by employees as a guideline in executing their tasks. As any change is usually unwelcome, every effort should be made to motivate individuals to acquire any new skills that may be needed.

A major shift in company strategy may demand a restructuring of the organisation. The commonest change is an expansion of business activities which will place a greater workload on most departments. To accommodate this new strategy, additional staff will be recruited and departmental managers may create specialised subdepartments to sustain an effective level of operation. When the firm has outgrown a single product line and becomes a multi-product organisation, diversification will typically be accompanied by a move to a divisional structure. Operating decisions are handled by the division managers while strategic planning and co-ordination remains under the control of head office. As the complexity of the diversification process continues, divisions will tend to acquire control over short to medium-term management issues, thereby permitting head office to concentrate on strategic planning for the longer term. This is known as a decentralised management structure. The risk in this type of structure is that head office may lose control and that incorrect decisions at divisional level may place the financial future of the firm at risk. It is therefore mandatory that decentralisation is accompanied by the creation of an adequate system of control through which head office can monitor divisional performance.

Early research on organisation structure concluded that there is always a 'best way' to design organisations to suit specific strategic development. More recently, however, there has been growing acceptance of the 'contingency theory' which suggests that there is no single structure. Instead it is suggested that design must be capable of allowing the firm to carry out the tasks to be performed and to fit the environmental or technological variables which can influence corporate growth objectives. The implications of the contingency theory is that relatively static, centralised structures can handle stable, predictable conditions. Under dynamic or uncertain conditions, however decentralised, flexible approaches are more appropriate.

Companies in the aquaculture sector face highly uncertain conditions because of the emphasis on introducing new technology to solve culture problems, stabilise yields and reduce production costs. This is reflected by highly flexible management structures which, at the extreme, may result in the abolition of departmental barriers with all the employees contributing to the solution of research problems. As an aquaculture company resolves the important culture issues and moves towards the establishment of a standardised production process, this will be paralleled by the creation of a more formal, departmentalised structure. Nevertheless the lessons learned during the more uncertain period while evolving a commercially viable culture process often results in management retaining a more flexible approach to decision making and responsibility execution. The degree of flexibility will vary by department, however, with those managers in areas of accounting or shipping tending to place emphasis on standard procedures to control their employees. In contrast, companies in the more traditional sectors of the industry rely heavily on highly structured departments in all areas of the company operating under rigidly enforced procedures. The senior management in these companies consider that without strict policies, there would be the risk of mistakes being made. On the other hand, this viewpoint is possibly stifling innovation to the point where it creates a totally stagnant management environment, and the firm's ability to react to change or identify new opportunities for profitable growth may be totally vitiated. Given the poor performance in recent years of the larger, long-established companies in the catching and processing sectors, it is reasonable to speculate that these business failures were in part caused by an excessively rigid style of management.

Management researchers in other industries have identified a close correlation between a high degree of flexibility in organisational style and a firm's capability to succeed in the post-war business environment. There is no reason to believe that the relationship is any different in the fishing industry. The introduction of this more flexible management style can only be initiated by the chief executive either through the employment of training consultants to instruct existing personnel or by the recruitment of new management staff.

CONTROL SYSTEMS

There is little benefit from developing and implementing new strategies if the results are not evaluated. Any strategic change should be accompanied by the creation of a control system to assess performance. The basic elements of this system are the setting of standards of performance, the evaluation of actual results against the standards, and the capability to introduce corrective actions when necessary.

Any business activity will generate information, and the objective of a control system is to create a feedback by which data on performance can be re-applied to the information base created during the strategy development process. A simple example of the feedback structure is given in Figure 4.1. The possible cause of a re-evaluation may be an adverse variance in actual versus forecasted performance or because new information indicates a need to re-assess the logic underlying the strategic planning process.

Although the control system indicates a feedback to every level of the planning process, it is more likely that revisions will be made at the forecast or business plan stage than that a variance in performance will require restatement of the overall corporate objective or strategy. A more

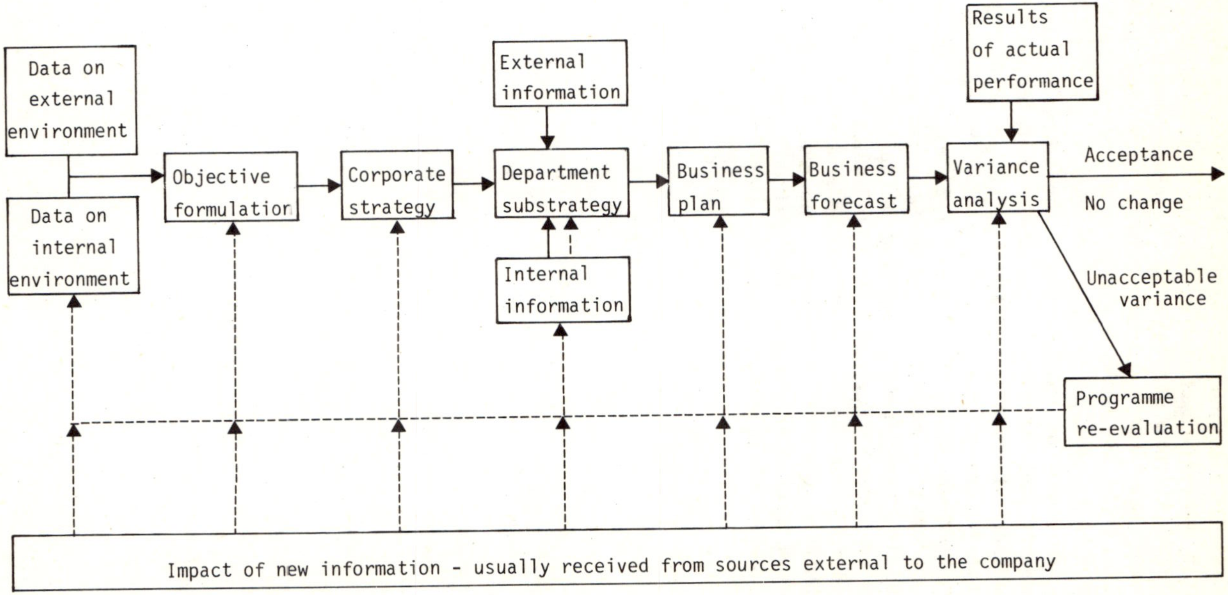

Fig 4.1 The feedback elements within the strategic management process to accommodate the impact of information

likely cause of such a fundamental revision would be the impact of new information concerning a variable of important influence on future performance (eg the announcement of a new tariff barrier by the country which currently provides over 50% of a firm's export sales).

An effective control system can be used to anticipate future problems as early trends in actual versus planned performance can provide sufficient warning to allow the introduction of remedial plans before adverse results cause financial damage to the organisation. The control system is also recognised as a model of the firm, and in some industries, operations researchers have evolved sets of equations which assist managers in deciding on the optimum solution to changes in the value of key influence variables. This approach is probably one of the most powerful tools available to modern management. To date, however, there is little indication that the concept has gained more than a minimal acceptance in the fishing industry.

5
Accounting methods

DOUBLE-ENTRY ACCOUNTING

The purpose of an accounting system is to provide a record of all financial transactions. In most countries there are specified standards to ensure a commonality of approach to enable anyone, whether inside or outside an organisation, to understand these financial records. The objective of the record keeping is to describe the process of exchange, and this can be achieved by using a double-entry system in which both aspects of exchange - giving and receiving - are recorded.

Each account within the system has a debit side in which items received are described, and a credit side to represent those given. A simple illustration is provided by a fish wholesaler who purchases product at auction for cash. The purchase account would record the value of fish purchased as follows:

Fish Purchase Account

	Debit	Credit
January 1	£100	

Every entry in the double-entry system must be reflected by a second entry in another account. In this example, there is a 'receiving' or debit entry for £100 that must be matched by a 'giving' or credit entry, which in this case would be shown as follows:

Cash Account

	Debit	Credit
January 1		Fish purchase £100

FINANCIAL RECORDS

On a regular basis, usually every twelve months, it is necessary to summarise the credit and debit entries in a form which provides a financial record of the organisation's accounting transactions. To further illustrate the double-entry method and the preparation of financial records, the following activities for a thirty day period by Mr Nicholas, the owner of an in-shore trawler, can be used:

(1) He owns a trawler valued at £30 000 and equipment of £10 000. There is £7 000 in the business current account and the bank loaned the business £32 000 to purchase the boat and the equipment.
(2) On 2 June, £400 of fuel oil was purchased on credit from the oil company.
(3) On 3 June, £200 of boat supplies were purchased for cash.
(4) On 5 June, £300 of catch was sold at auction for cash.
(5) On 8 June, £600 of catch was sold on credit to a fish processor, Smiths Ltd.
(6) On 11 June, additional boat supplies were purchased for £100 cash.
(7) On 13 June, £500 of catch was sold for cash at auction.
(8) On 16 June, £400 of catch was sold for cash at auction.
(9) On 19 June, £500 of catch was sold for cash at auction.
(10) On 20 June, additional boat supplies were purchased for £300 cash.
(11) On 22 June, £400 of catch was sold on credit to Smiths.
(12) On 26 June, £500 in cash was paid for repairs to the winch and generator.
(13) On 30 June, wages of £250 were paid to the crew.
(14) On 30 June, £200 was paid as the monthly interest charge on the bank loan.
(15) On 30 June, payment of £1 000 was received from Smiths.
(16) On 30 June, the oil company bill of £400 was paid.

Each of these transactions must be recorded as a credit and a debit in the relevant accounts in a double-entry system. The Capital account records all transactions between the proprietor and the business. A credit balance indicates the amount owed by the business to the proprietor, whereas a debit balance indicates the amount owed by the proprietor to the business. At the beginning of June, the Capital account will be in credit for the amount equal to the sum of 'assets' (ie property owned by the business: in this example cash +

boat + equipment) of £47 000, less 'liabilities' (ie debts owed by the company: in this case the bank loan) of £32 000 which equals £15 000. The assets are shown as opening debit entries and the liabilities as a credit entry. Thus the opening balances will be as follows:

<u>Capital Account</u>

Debit	Credit
	June 1 Opening bl. £15 000

<u>Bank Loan</u>

Debit	Credit
	June 1 Opening bl. £32 000

<u>Cash</u>

Debit	Credit
June 1 Open. bl. £ 7 000	

<u>Trawler</u>

Debit	Credit
June 1 Open. bl. £30 000	

<u>Equipment</u>

Debit	Credit
June 1 Open. bl. £10 000	

The purchase of fuel involves the receipt of the purchase by the business and the creation of a debt with the oil company. Hence the entries will be:

<u>Fuel Oil Account</u>

Debit	Credit
June 2 Oil comp. a/c £400	

<u>Oil Company</u>

Debit	Credit
	June 2 Fuel oil a/c £400

The purchase of boat supplies is recorded by:

Boat Supplies Account

Debit	Credit
June 3 Cash a/c £200	

Cash

Debit	Credit
	June 3 Boat supp. a/c £200

The landings of fish and the sale at auction for cash is reflected by a series of entries:

Fish Sales Account

Debit	Credit	
	June 4 Cash a/c	£300
	June 13 Cash a/c	£500
	June 16 Cash a/c	£400
	June 19 Cash a/c	£500

Cash Account

Debit		Credit
June 4 Fish sales a/c	£300	
June 13 Fish sales a/c	£500	
June 16 Fish sales a/c	£400	
June 19 Fish sales a/c	£500	

The sale of product on credit will require the same credit entry for the fish sales account, but the debit entry will be recorded as follows:

Fish Sales Account

Debit	Credit	
	June 8 Smiths a/c	£600
	June 22 Smiths a/c	£400

Smiths Ltd Account

Debit		Credit
Jur 8 Fish sales a/c	£600	
June 22 Fish sales a/c	£400	

The boat supplies purchases are as:

Boat Supplies Account

Debit			Credit	
June 11 Cash a/c	£100			
June 20 Cash a/c	£300			

Cash Account

Debit		Credit	
		June 11 Boat supp. a/c	£100
		June 20 Boat supp. a/c	£300

Similar entries are made for the winch and generator repairs:

Boat Repairs Account

Debit		Credit	
June 26 Cash a/c	£500		

Cash Account

Debit		Credit	
		June 26 Boat rep. a/c	£500

The expenses paid at the end of the month are recorded by:

Wages Account

Debit		Credit	
June 30 Cash a/c	£250		

Bank Interest Account

Debit		Credit	
June 30 Cash a/c	£200		

Oil Company Account

Debit		Credit	
June 30 Cash a/c	£400		

Cash Account

Debit		Credit	
		June 30 Wages a/c	£250
		June 30 Bank Int. a/c	£200
		June 30 Oil co. a/c	£400

The receipt of payment from Smiths Ltd. is described by:

Smith Ltd Account

Debit	Credit	
	June 30 Cash a/c	£1 000

Cash Account

Debit		Credit
June 30 Smith a/c	£1 000	

At this stage all the transactions for the period have been recorded in the appropriate accounts, and the next step is to use these data to prepare a summary of trading activity, accompanied by a statement of the current asset and liability situation. Trading performance is described by the construction of a Profit and Loss Account, while assets and liabilities are summarised in a Balance Sheet.

Before the preparation of these summaries, it is usual to execute a Trial Balance to check that each debit is matched by an appropriate credit entry. If all entries have been made correctly, the total debit entries will equal that of the credits and the accounts are said to be 'in balance'. The Trial Balance is constructed by listing the net balances for all the accounts, the balance in each account being the amount by which one side (credit or debit) exceeds the other. For example the total entries for Mr Nicholas' cash account are:

Cash Account

Debit		Credit	
June 1 Open. bal.	£7 000	June 3 Boat supp. a/c	£200
June 4 Fish sal. a/c	£300	June 11 Boat supp. a/c	£100
June 13 Fish sal. a/c	£500	June 20 Boat supp. a/c	£300
June 16 Fish sal. a/c	£400	June 26 Boat rep. a/c	£500
June 19 Fish sal. a/c	£500	June 30 Wages a/c	£250
June 30 Smiths a/c	£1 000	June 30 Bank int. a/c	£200
		June 30 Oil co. a/c	£200
	£9 700		£1 950

£7 750 Debit balance

The trial balance for the example trawler operation is:

Cash	£ 7 750	
Capital account		£15 000
Bank loan account		£32 000
Trawler account	£30 000	
Equipment account	£10 000	
Fuel oil account	£ 400	
Boat supplies account	£ 600	
Fish sales account		£ 2 700
Boat repairs account	£ 500	
Wages account	£ 250	
Bank interest account	£ 200	
	£49 700	£49 700

It must be stressed that the trial balance is not an integral part of the double-entry system; it is merely a method to determine whether the accounts are in balance with debits matched by credits. The above trial balance has a total of £49 700 of debit entries equalled by the same amount of credit entries, hence it can be assumed that the June entries have been made correctly.

The next step is the preparation of final accounts by transferring the various balances to the profit and loss account or the balance sheet. Accounts with a debit balance which represents a value which remains in the business (eg the trawler account) will appear - or more correctly 'are credited' - in the balance sheet. Any other debit balances must be expenses which are transferred to the profit and loss account. Those credit balances recorded in the accounts which represent an obligation to refund a value to an outside party (including the owner) will be transferred - 'or debited' - to the balance sheet. The credit balance of £32 000 in the bank loan account is of this type. All other credit entries not representing an obligation will be transferred to the profit and loss account.

Thus, using the information in the trawler operation example, the final summary of performance and the asset/liability status would be as follows:

Trading and Profit and Loss Account for the period ending June 30

	£		£
Fuel expense	400	Fish sales	2 700
Boat supp. exp.	600		

Boat reps. exp.	500			
Wages	250			
Bank interest	200			
Total expense		1 950		
Balance, being net profit transferred to the capital account		750		
		2 700		2 700

Balance Sheet as at June 30

Capital			*Fixed Assets*		
As at June 1st	15 000		Trawler	30 000	
Net profit June	750		Equipment	10 000	
		15 750			40 000
Current Liabilities			*Current Assets*		
Bank loan		32 000	Cash		7 750
		47 750			47 750

The profit and loss account is the summary document which will reveal whether sales revenue for the period exceeds expense thereby resulting in the business generating a profit. This profit is shown as a debit entry in the profit and loss account and is credited to the balance sheet as an obligation of value owing to the owner of the business. The liability to the owner along with the earlier profits and the capital invested, is the last debt that would be paid upon dissolution of the business. It therefore has a high degree of permanence and is shown by convention at the top of the balance sheet, to be followed in descending order of permanence by all other liabilities. The liabilities for which the most immediate repayment is required (eg money owed to suppliers, a bank overdraft) are shown as current liabilities. The same conventions are used for assets, with the most permanent being shown first as fixed assets, followed by those assets which could more readily be converted into - or are already in the form of - cash. The latter are known as current assets.

The example accounts are for a sole trader; other common types of business structure are partnerships and limited liability companies. The limited company is created by the issue of shares, and its advantage over other business types is that the shareholders are only liable for payment of debts to the value of their investment in shares. In the case of sole traders or partnerships, the owners are not able to restrict the liability for debts and their personal estates can be called upon by creditors. Limited companies have a number of other advantages related to their ability to raise additional capital and to sustain the permanence of the entity from one generation to the next. The differences in legal status will mean that there are variations in the way the accounting records are presented. For example in the balance sheet of a limited company, the capital (ie shares issued) must be kept distinct from profits earned, whereas in the sole trader trawler example earlier, the profits were combined with existing balance in the capital account. It should be emphasised, however, that despite some differences in the recording of transactions, the fundamental concepts of debit and credit are the same for all types of business entity.

In presenting the concept of financial records in the above example, certain simplications were made for the sake of clarity, namely:

(1) All goods sold were disposed of during the accounting period.
(2) All incurred expenses and expense payments made relate to activities which occurred during the accounting period.
(3) Fixed assets suffered no loss in value during usage.
(4) The business received all payments due from customers in the accounting period.

These assumptions are rarely true in genuine business situations and it is therefore necessary to understand the impact of their removal.

Virtually all trading concerns will maintain a stock of goods to meet the day-to-day requirements of their customers. At the beginning of the accounting period, goods on hand will be shown on the balance sheet as an asset. During the accounting period additional goods are acquired or manufactured for sale to customers. Cost of goods sold can be calculated from the formula:

(Goods on hand at the beginning of the period +
Goods acquired or manufactured during the period) -
(Goods on hand at the end of the period)

Thus a wholesaler who starts a period with £5 000 of fish, acquires £15 000 of fish during the period and has £3 000 of fish on hand at period end, can calculate cost of goods sold as (£5 000 + £15 000) - (£3 000) which equals £17 000.

Many of the incurred expenses will not exactly match the accounting period, yet full payment may have been made. When this occurs, that part of the payment which relates to the period will appear in the profit and loss account as an expense. The balance is a payment in advance for future benefit and will appear as an asset in the balance sheet. If, for example, a business contracted to rent warehouse space for twelve months half way through the financial year for £1 000, the relevant entries would be credit cash £1 000, debit rent account £1 000. At year end, £500 would be credited as an expense to the profit and loss account. The balance of £500 will be shown in the balance sheet as an asset. In the case where expense is incurred by payment not made (eg the purchase of raw materials on credit), then the money owed by the business to the supplier will appear as a current liability in the balance sheet.

The use of a fixed asset such as plant or machinery during the accounting period will cause 'wear and tear' and thereby reduce the value of the asset. This effect is recorded through the mechanism of depreciation by a debit to the profit and loss account to reflect the expense of asset usage. An adjustment to the value of the asset is achieved by a credit entry against the appropriate fixed asset in the balance sheet. As it is difficult to define the exact loss in value from usage, the usual convention is to apply a standard charge for depreciation in direct proportion to the original cost of the asset. There are a number of formulae for calculating depreciation, and probably the simplest to understand are 'straight line' and 'declining balance'. With the former, the cost of the asset is divided by the predicted number of years of its useful life and the resultant figure deducted every year. On this basis, an asset which cost £20 000 and estimated to have a ten year life, would be depreciated by £2 000 every year (ie £20 000 ÷ 10). Using the declining balance formula, a constant percentage of the value is used to calculate depreciation. Thus if a machine cost £10 000, and a 10% depreciation rate selected, then in year one the depreciation cost would be £1 000 (ie £10 000 x 0.10). In year two the value of the asset is equal to the original value less the prior year depreciation expense (ie £10 000 - £1 000 = £9 000) and the depreciation in the second year will be £900 (ie £9 000 x 0.10). This means that the earlier years of an asset's life are charged with a higher rate of depreciation, which is reasonable as most assets decline in efficiency with age.

The effect of all these accounting concepts on year-end accounts can be illustrated by preparing final accounts for a small company which processes frozen, breaded fillets. The trial balance as at 31 December is as follows:

		Debit £	Credit £
Creditors			30 000
Debtors		34 000	
Stock at Jan 1st			
Raw materials	6 000		
Work in progress	8 000		
Finished goods	10 000		
		24 000	
Cash at bank - overdrawn			3 000
Discounts received			200
Discounts allowed		500	
Sales			214 700
Bad debt provision Jan 1st			1 000
Factory wages		39 000	
Administrative expense		26 000	
Bad debts		800	
Interest paid		3 000	
Factory expense		22 500	
Selling expense		15 700	
Provision for depreciation as at Jan 1st			60 000
General reserve			28 000
Plant and machinery at cost		160 000	
Raw material purchases		60 000	
Issued share capital			160 000
Profit and loss balance as at Jan 1st			40 000
Freehold land and building at cost		151 400	
		536 900	536 900

At year-end the following facts and decision will have to be accommodated in the preparation of final accounts:

(1) The stocks at 31 December are raw materials £8 000, work-in-progress £10 000, finished goods £8 800, or in total stocks of £26 800.
(2) The decision on dividends is that a total payment be made to shareholders of £12 000.
(3) The provision for bad debts, which is an estimate of the possible money owed to the company that will not be paid, is to be set at 5% of debtors.
(4) £20 000 is to be transferred to the general reserve.
(5) Warehouse rent for the next six months was paid on 1 October.
(6) It is estimated that fuel oil to the value of £100 has been consumed but not yet billed by the supplier.
(7) Depreciation of plant and machinery at 10% of cost.

Incorporation of these facts into the final accounts would produce the following profit and loss account and balance sheet:

Trading and Profit and Loss Account for the year ended December 31

		£		£
Purchase	60 000		Sales	214 700
Stock at Jan 1	24 000			
		84 000		
Less stocks as at Dec 31		(26 800)		
Factory wages		39 000		
Factory expense		22 500		
Cost of goods sold		118 700		
Gross profit c/d		96 000		
		214 700		214 700
Administrative expense*		25 500	Gross profit b/d	96 000
Selling expense		15 700	Discounts received	200
Bad debts**		1 500		
Discounts allowed		500		
Interest paid		3 000		
Depreciation***		16 000		
Net profit c/d		34 000		
		96 200		96 200
Dividends for year		12 000	Net profit b/d	34 000
General reserve		20 000	Balance from last year	40 000
Balance c/d		42 000		
		74 000		74 000

* Administrative expense is composed of the account balance of £26 000 to which is added the fuel oil used (£100) not yet billed by the supplier. It is also necessary to deduct that part of the warehouse rent paid but which applies to the following year. This is calculated by dividing the rent £1 200 by 2 to yield £600. Hence the adjustments to administrative expense = £26 000 + £100 (oil) - £600 (rent) = £25 500.

** The bad debt provision of 5% of debtors equals £1 700 (ie £34 000 x 0.05). As at 1 January, the provision stood at £1 000 and hence an increase in provision of £700 (ie £1 700 - £1 000) is needed. This additional expense when added to the actual bad debts during the year of £800, yields a total expense of £1 500.

*** The cost of plant and machinery is £160 000 and therefore depreciation is calculated at £160 000 x 0.10 or £16 000.

Balance Sheet as at December 31

Share Capital and Reserves	£	*Fixed Assets*			£
Share capital	160 000	Land and buildings			151 400
General reserve*	48 000	Plant and machinery			
Profit and loss acc.	42 000	Cost		160 000	
		**Acc dep		76 000	
					84 000
					235 400
		Current Assets			
		Stocks		26 800	
		Debtors	34 000		
		less prov	1 700		
				32 300	
		Rent in adv		600	
				59 700	
		Less Current Liabilities			
		Creditors	30 000		
		Fuel oil exp	100		
		Dividend due	12 000		
		Overdraft	3 000	45 100	
					14 600
	250 000				250 000

* The general reserve is calculated from the balance at 1 January of £28 000 to which is added at year-end another £20 000.

** Accumulated depreciation represents accrued depreciation at 1 January of £60 000 plus that for the current year of another £16 000 as shown in the profit and loss account.

6
Cost accounting

THE COMPONENTS OF COST

The generation of profit is dependent upon sales revenue exceeding costs. Revenue is controlled by two variables, price per unit and total units sold. The components of cost, however, tend to be more complex. For a company which makes no modification to products before resale (eg a port wholesaler who buys fish at auction which is then sold to inland wholesalers) the major cost components are the expenses associated with buying, marketing, storage, delivery and general administration. A processor (eg a sardine canner) faces the same costs as the wholesaler, but in addition will incur other expenses during processing. If either type of business is to succeed, it is vital that management have a full understanding of the cost elements associated with the product transformation and transaction activities.

Costs can be divided into two main types, variable and fixed. The former vary in direct proportion to the level of business activity (eg the fuel consumed by a trawler out fishing). Fixed costs are those expenses which remain unchanged despite variations in the level of activity (eg the interest payments on a trawler which are due even when the boat is tied up during bad weather). Total costs of the firm are based upon the combined value for fixed and variable expense. In order to reach a decision on output, it is important to establish the point at which sales revenue equals total costs, known as the 'break even' point. This is important because at an output level beyond break even, the firm will generate profit. Break even can be evaluated graphically by plotting revenue and costs against output or alternatively break even in units of output can be calculated from the formula

Break even = (fixed costs + zero profit) ÷ (unit price - unit variable costs)

The formula could be used by a trout farm whose fixed costs for the construction of the facility is £20 000 and the estimated costs for labour, feed and miscellaneous items is £0.50 per pound of trout. The current market price ex-pond for fresh trout is £0.90/pound. The break even output in pounds of fish

= (£20 000 + £0.00) ÷ (£0.90 - £0.50)
= 50 000 lbs of fish

The calculation can be checked to see if profit is zero at the break even output level. Sales revenue equals £45 000 (ie 50 000 x £0.90) as does total cost [ie £20 000 + (50 000 x £0.50)]. Hence profit at this output level is zero and therefore the break even point has been correctly estimated.

PRODUCTION COST ANALYSIS

Although all costs should be understood, the most critical area is probably that of production. Companies in the fishing industry involved in this activity include processors (eg canners, fillet producers, smokers) and virtually all of the firms in the aquaculture sector. It is usual to express expense in relation to units of output in order to understand costs. The simplest situation occurs in companies engaged in 'job production' where the units of output per year are small. A company which builds trawlers, for example, may have the following production costs:

Direct Costs	£
Labour	38 400
Materials	96 000
Overhead Costs	
Office administration	12 000
Lighting, heating	8 000
Rent, taxes, depreciation	6 000
Total costs	160 400
Boats produced	5

The breakdown of costs into direct and overhead is based upon the fact that direct costs are those specifically attributable to boat construction. Overhead costs are those which cannot be allocated to a specific part of the construction programme. In some companies, overhead costs tend to be fixed and will remain relatively unchanged despite variations in output. The cost data can be used to estimate average costs per trawler built by dividing costs by output to yield the following:

Direct cost/trawler	=	£134 400 ÷ 5	=	£26 880
Overhead cost/trawler	=	£ 26 000 ÷ 5	=	£ 5 200
Total cost/trawler	=	£160 400 ÷ 5	=	£32 080

In many industries, a large number of similar units are manufactured using mass production methods. The same concept of estimating unit cost is applied as in job costing, except that the output is numbered in many thousands. This type of evaluation is known as 'process costing' and can be demonstrated by examining the production data for a processor who manufactures three forms of coated, frozen products:

Cost component	*Batter portions*	*Breaded portions*	*Fish sticks*
Fish block costs	82 944	68 424	45 624
Batter, breadings	9 216	4 608	3 072
Packaging	7 680	4 608	3 072
Direct labour	42 240	17 280	17 280
Total direct costs	142 080	94 920	69 048
Production pounds	384 000	203 400	153 600
*Labour hours**	21 120	10 368	6 912

* Calculated from the formula labour hours = total employees x hours worked per employee.

The direct cost per pound of output for the three items can be calculated by dividing total costs by production output which yields a direct cost per pound of £0.37 for batter portions, £0.47 for breaded portions and £0.45 for fish sticks. Other costs involved in the production operation are indirect labour (ie employee hours which cannot be allocated to a specific product for such staff as production engineers and quality control technicians) of £36 000, heating and lighting at £43 200, equipment leases and rentals £45 600 and plant depreciation of £24 000. In total these overheads represent an additional annual cost of £148 800.

To establish the total production cost per unit of output it is necessary to allocate the overhead costs to each unit of finished product. How this is achieved is often a matter for debate and financial advisors will frequently disagree over the allocation technique to be used in a specific situation. In general, however, the objective is to allocate overheads in a way which equitably reflects the level of production activity. One common approach is to assign a fixed charge per unit of output, calculated on the basis:

Fixed overhead per unit = (total overhead costs) ÷ (total units of output)

Hence in the above example this formula would yield an overhead per pound of output of £0.2008 (ie £148 800 ÷ 741,000). When this is added to the direct costs, it yields a cost per pound of output of £0.5708 for batter portions, £0.6708 for breaded portions and £0.6508 for fish sticks.

The allocation of overhead by unit of output can be criticised on the grounds that the product with the largest volume of output receives the greatest allocation of total costs. It is argued that a more accurate allocation of costs is that done on the basis of labour hours. In this way the product which consumes the most labour hours will receive the majority of overhead costs. This encourages the introduction of revised manufacturing methods which can reduce the labour input for a product and hence the cost per unit of output. Allocation by labour hours is calculated with the formula:

Overhead cost per labour hour = (total overhead costs) ÷ (total direct labour hours)

which in the example would yield £3.875 per labour hour (ie £148 800 ÷ 38,400). The cost allocation per unit of output can be estimated from the formula:

Allocation per output unit = [(direct labour hours) x (cost/hour)] ÷ (total output units of product)

which yields for the three products, overhead costs per pound of £0.213, £0.198 and £0.174 respectively. When added to direct costs this results in a total cost per pound of £0.583 for batter portions, £0.668 for breaded portions and £0.624 for fish sticks. Compared with the previous method of allocation by volume of output, allocation by labour hours causes an increase in costs for batter portions (ie from £0.5708 to £0.583) due to the high labour content associated with the production process. This change may seem marginal on the basis of cost per unit of output, but when expressed as a total cost for the year, the increase of £0.0122/pound means that total costs for batter will be increased by £4 684.80 over twelve months.

STANDARD COSTS

Although it is necessary to understand actual costs in order to make price or profit margin decisions, such cost data are of little use in evaluating the efficiency of the manufacturing process unless they can be compared with predetermined values. These are usually referred to as 'standard costs' and represent costs incurred when the processing facility is operating at an optimum level. By comparing the actual costs with the standard, the difference or 'variance' will reveal which aspect of the production process is the cause of adverse performance.

A cost is typically comprised of two components, the quantity and cost of ingredients. In the case of standard costs the quantities are often expressed as a proportionate constant of the volume or weight of finished goods. For the fish processor example the standard costs might be presented as follows:

Component	*Batter portion*	*Breaded portion*	*Fish sticks*
Pounds/shift finished product	3 200	3 200	3 200
Pounds/shift fish blocks	1 280	1 888	1 920
Cost/lb fish blocks	£0.54	£0.50	£0.50
Pounds/shift batter, breading	1 920	1 312	1 280
Cost/lb batter, breading	£0.04	£0.05	£0.05
Packaging/lb finished product	£0.02	£0.02	£0.02

Labour hours/shift	176	144	144
Labour hour cost	£2.00	£2.00	£2.00
Overhead cost/per labour hour	£3.875	£3.875	£3.875
Overhead allocation/ shift	£682.00	£558.00	£558.00
Output cost per shift	£1 866.00	£1 919.60	£1 934.00

The first step in variance analysis is to compute total variance and then subdivide the variance into the components of price (or rate) and usage (or efficiency). Efficient operating control is reflected in the objective of maximising output using a given quantity of inputs or, expressed another way, utilising a minimum level of inputs to produce a given quantity of output. This perspective on input-output is important in understanding the principles underlying the concepts of variance analysis.

The aim of the manufacturing department is to produce finished goods at standard cost. The standard for any component of finished goods, whether materials or labour, is equal to (the quantity of finished goods) x (the standard proportion of the input component per unit of finished goods) x (the standard cost per unit of the input component). Thus for an output of 9 600 lbs of batter portions in the above example, the standard input cost for fish blocks would be £2 073.60 (ie 9 600 x 0.40 x £0.54). Whereas the actual costs for any input component is equal to (the actual input quantity used) x (the unit cost of input). Hence if a production run of batter portions used 4 032 lbs of fish blocks at £0.55/lb, then the actual input cost would be £2 217.60.

Total variance is the difference between standard cost of the input component for a given quantity of output using actual unit input cost for the component. In the batter portion example, this difference is -£144.00 (ie standard cost of £2 073.60 minus the actual cost of £2 217.60). As actual costs in this situation exceed the standard cost, the variance is negative and described as 'unfavourable'. Where actual costs are less than standard, the variance would be positive or 'favourable'.

To isolate the impact on total variance of quantity and costs of inputs it is necessary to calculate 'unit cost (or price) variance' which equals:

(Actual unit cost - standard unit cost) x (actual input quantity used)

and the 'quantity (or usage) variance,' which equals:

(Actual input quantity - standard input quantity) x (standard unit cost)

When applied to examine the variance components of total unfavourable variance of £144.00 for batter fish portions, the formulae yield:

Unit cost variance	= (£0.55 - £0.54)(4 032)	=	£ 40.32U
Quantity variance	= (4 032 - 3 840)(£0.54)	=	£103.68U
Sum of variances	= (£40.32) + (£103.68)	=	£144.00U

From this analysis it can be concluded that usage variance at £103.68 is the most important cause of unfavourable variance, contributing 72% of the total problem.

It is important to isolate the effect of cost and quantity variance because, for example, if one is favourable and the other unfavourable, their combined effect may lead the production to assume the operation is meeting predetermined standards in all areas. This can be illustrated by examining the direct labour costs for the production of 4.5 shifts of fish sticks. The standard cost for the labour component is equal to (4.5) x (144) x (£2.00) or £1 296. The actual direct labour costs according to the wages account were £1 302.72 or an adverse variance of only +£6.72 which would appear of little concern. Further analysis revealed, however, that actual input hours was 708 and the average labour rate paid per hour was £1.84. Hence component variance analysis yields:

Unit cost variance	= (£1.84 - £2.00)(708)	=	-£113.28F
Quantity variance	= (708 - 648)(£2.00)	=	£120.00U
Sum of variances	= (-£113.28) + (£120)	=	£6.72U

which reveals that although labour hour costs have been kept below standard, the gain from this action is more than offset by a major increase in total direct labour hours well above standard.

OVERHEAD ALLOCATION AND ABSORPTION

As described earlier, the majority of overhead costs are fixed and are incurred without relation to actual production activity. In order to reflect the overheads as a component of cost in the finished goods, therefore, it is necessary to allocate overhead at a predetermined rate which is related to some form of measurable activity. In the fish processor example it was set at £3.875 per direct labour hour and was calculated from the formula:

Overhead allocation = (budgeted total fixed costs) ÷ (budgeted total direct labour hours)

The overhead allocation to finished product is then assigned on the basis of:

(total direct labour hours/product)(overhead allocation rate)

Thus if the production of one shift of breaded portions involved 144 direct labour hours, the overhead allocation would be £558 (ie (144)(£3.875)). This allocation is usually referred to as 'absorption' of overhead.

Overhead absorption poses no problem as long as the business is operating on budget. However, if actual production is at variance with the budget, the finished goods will not correctly reflect the firm's fixed costs. In the fish processor example, a 20% reduction in production volume would be reflected by a fall in direct labour hours from the budgeted figure of 38 400 to an actual of 30 720 hours. The resultant overhead allocation would be £119 040 (ie (30 720)(£3.875)). At this level the allocation would be £29 260 below the budgeted figure of £148 880 and finished goods at standard cost would therefore understate the true costs by this amount. Conversely, if output exceeds budget by 20%, total direct labour hours would rise to 46 080 hours and the allocated overhead would be £176 560 (ie (46 080)(£3.875)). This would be reflected in finished goods and hence these costs would be overstated by £29 760. The 'overabsorption' of true costs would then cause gross profit to be understated by this amount.

With overhead variance analysis, the rate of cost absorption complicates the evaluation. For example, the accounting department at the fish processor has produced information on actual total fixed costs as being £156 800, whereas the amount allocated to finished goods was £141 422 indicating an unfavourable variance of £15 378. This first component of variance can be assessed by examining total costs versus budget for the elements which contribute to total fixed costs.

	Budget(£)	*Actual(£)*	*Variance(£)*
Indirect labour costs	36 000	42 000	6 000 Unfavourable
Heating, lighting	43 200	47 800	4 600 Unfavourable
Leases, rentals etc	45 600	43 000	2 600 Favourable
Depreciation	24 000	24 000	0 No variance
Total fixed costs	148 800	156 800	8 000 Unfavourable

The overhead allocation to finished goods consists of:

	Batter	*Breaded*	*Sticks*	*Total*
Total direct hours	19 360	9 936	7 200	36 496
Overhead rate/hour	£3.875	£3.875	£3.875	£3.875
Overhead allocation(£)	75 020	38 502	27 900	141 422

and activity variance can be estimated from the formula:

(actual direct labour hours - budgeted direct labour hours) x (overhead standard rate)

which in this example

= (36 496 - 38 400) x (£3.875) = -£7 378

Thus of the total overhead variance of £15 378, £8 000 or 52% is caused by actual costs exceeding budget and the balance of £7 378 by underallocation of overhead to finished goods because production volume was below forecast.

Although fixed costs and overhead rates are important for both product costing and pricing purposes, such rates have limited significance for control purposes because there is little opportunity for the production supervisors to influence the issues of depreciation, leases and taxes.

These factors are determined by the accounting department or organisations outside the company. Hence improvements in manufacturing efficiency will have no direct impact on fixed costs other than to provide a reduction in variable costs which may act as an offset against rising overheads. This weakness in the overhead allocation concept has caused some firms to move to a method known as the 'contribution' approach in which only the costs allocated to finished goods are those which are truly variable in nature. Fixed costs are then reflected as an adjustment to the profit and loss account. The concept can be illustrated by the fish processor example, initially by assuming sales and production are on forecast.

Comparative Profit and Loss Accounts(£)

Method	*Overhead*	*Contribution*	*(Difference)*
Sales	649 783	649 783	
Cost of goods	454 848	306 048	-148 800
Gross profit	194 935	343 735	+148 800
Administrative exp	129 800	129 800	0
Fixed expense	N/A	148 800	+148 800
Net profit	65 135	65 135	0

Under the circumstances of the firm operating on budget, the two approaches generate the same net profit. When actual performance is at variance with the budget, the contribution method provides a truer picture of financial performance because the actual fixed costs are deducted from gross profit, no matter what the output. The impact of this effect can be shown for the fish processor when sales are above budget by 20%:

Comparative Profit and Loss Accounts(£)

Method	*Overhead*	*Contribution*	*(Difference)*
Sales	779 740	779 740	0
Cost of goods	545 818	367 258	-178 560
Gross profit	233 922	412 482	+178 560
Administrative exp	129 800	129 800	0
Fixed expense	N/A	148 800	+148 800
Net profit	104 122	133 882	+ 29 760

and when sales are below budget by 20%:

Method	*Overhead*	*Contribution*	*(Difference)*
Sales	519 826	519 826	0
Cost of goods	363 878	244 838	-119 040
Gross profit	155 948	274 988	+119 040
Administrative exp	129 800	129 800	0
Fixed expense	N/A	148 800	+148 800
Net profit	26 148	(3 612)	- 29 760

When sales are over budget, overhead allocation results in an overabsorption of overhead and an understatement of net profit of £29 760. The reverse occurs when sales are below budget, because an under absorption of overhead using the allocation approach results in an overstatement of net profit of £29 760 for the year.

7
Investment appraisal and financial analysis

INVESTMENT APPRAISAL

The attainment of corporate objectives, such as growth in profitability, is to a great extent dependent upon investment in new projects which are capable of generating incremental revenue. Without such investment, the firm will inevitably approach the point at which the assets can no longer support a competitive position in the market. Failure to reverse this trend will eventually lead to business collapse.

No company has unlimited access to the financial capital required to support new investment, and therefore it is important that any new project has the potential to maximise future profits. There are a number of techniques available for investment appraisal: their application can be illustrated by an evaluation of the financial implications of establishing a pond-based penaeid shrimp farm. The proposal assumes the creation of a farm comprising a hatchery, algae production unit (for feeding the shrimp larvae in the hatchery), and a pond grow-out system capable of supporting a yield of 1 000 lbs per acre of heads-on shrimp which will process out as 61-70 count shell-on tails. The initial investment would be the construction of the farm over a twelve month period involving the following costs:

Hatchery	£
Building	61 500
Rearing tanks	13 700
Filter system and pumps	9 000
Emergency generator	5 000
Instruments	4 400
Compressor and refrigerators	2 900
Plumbing and wiring	4 200
Miscellaneous items	8 400
Total cost	109 100

Algae Facility	
Tanks, pumps and hoses	4 050
Lights and wiring	2 250
Centrifuges	10 000
Miscellaneous items	3 700
Total cost	20 000
Pound Grow-out System	
Land	130 000
Pond construction	173 020
Buildings	13 500
Machinery and equipment	34 400
Vehicles	12 000
Miscellaneous items	15 100
Total cost	378 020
Total investment	507 120

In year 1, the staff will be engaged in the supervision of the construction and evaluation of the final system through working with a pilot operation. During year 2, the system will be brought to 50% of the planned output level, and the final output level will be achieved in year 3. The total life of the facility is estimated at ten years, with the following revenue and operating cost profile:

	Year 1	*Year 2*	*Year 3*	*Year 4-10*	*Total*
Sales		157 633	315 265	2 206 855	2 679 753
Costs	74 467	113 552	152 636	1 068 452	1 409 107
Profit	-74 467	44 081	162 629	1 138 403	1 270 646

One traditional analysis technique is to calculate 'project payback', which is the time taken for the project to generate profit equal to the original investment. Because of the typically fluctuating nature of financial returns, payback is usually calculated using the formula:

Payback = (total initial investment) ÷ (average annual profit)

which in the shrimp farm example

= (£507 120) ÷ (£1 270 646 ÷ 10) = 3.99 years

Although the method is widely accepted, as an analysis technique it has limited use because it provides only a minimal description of the relationship between the cost of the assets and the profit generated. A more effective approach is to assess average annual Return On Investment as described in chapter 2. Applying this method to the farm example yields:

$$ROI = ((£1\ 270\ 646 \div 10) \div £507\ 120) \times 100 = 25.1\%$$

A criticism of ROI is that it does not reflect the effect of time on cash flows, ie profit generated in ten years time is assumed to have the same value as that received in earlier years. Yet given a choice between receiving £1 000 now or in a year's time, most people would accept the offer now. The £1 000 could be invested immediately and at a simple interest rate of 10%, would be worth £1 100 by the end of the year; or stated in another way, the value of the £1 000 if not accepted until a year from now is reduced by the interest foregone, ie £1 000 - £100 (the interest at 10%) = £900.

This concept of 'time is money' can be applied to investment appraisal by using discount factors which reflect the effect of 'waiting for money' and which express all future revenue or cost in the form of a present day value. The two accepted approaches are known as the 'present value' and 'Internal Rate of Return' (IRR) techniques.

The present value of future receipts can be estimated from the formula:

$$P = S \div (1 + i)^t \qquad (9:1)$$

where

P = the present value
S = the stated value of a sum in the future
i = the compound interest rate assumed
t = the time period before S is received

Thus the present value of £24 900 due in five years time at an assumed compound interest rate of 20% using equation 9:1 yields

$$24\ 900 \div (1 + 0.20)^5 = £10\ 000$$

For investment appraisal 9:1 can be used to evaluate whether the net present value of future revenue exceeds the present value of cash outflows (ie the capital and operating costs) using the formula Net Present Value (NPV)

$$= \sum_{t=0}^{n} A_t \div (1 + i)^t \qquad (9:2)$$

where

A_t = net cash flow in period t
t = the specified time period
i = the assumed interest cost

Application of equation 9:2 can be demonstrated by evaluating the NPV for a project involving an investment of £36 000 in four freezer trucks to open a new van sales operation. The profit yield is forecasted at £11 200 in year 1, £12 000 in year 2, £14 000 in year 3, £12 000 in year 4 and £10 000 in year 5. The assumed cost of borrowing is 20%. The net present value will be:

$$\frac{(-36\ 000)}{(1+0.2)^0}+\frac{(11\ 200)}{(1+0.2)^1}+\frac{(12\ 000)}{(1+0.2)^2}+\frac{(14\ 000)}{(1+0.2)^3}+\frac{(12\ 000)}{(1+0.2)^4}+\frac{(10\ 000)}{(1+0.2)^5}$$

This analysis could become quite tedious and therefore present value tables are published which give discount factors for £1.00 over various periods of time and levels of interest. The selected factor when multiplied by the future cash flow will yield the present value for the amount involved. By using this approach for the freezer truck example, the following result is achieved:

Year	*Value(£)*	*Discount factor*	*Present value(£)*
0	-36 000	1.000	-36 000
1	11 200	0.833	9 330
2	12 500	0.694	8 675
3	16 000	0.579	9 264
4	14 000	0.482	6 748
5	8 000	0.402	3 216
Net present value			+1 233

In this situation, the net profit exceeds the cash outflow for the original investment by £1 233 and therefore the project should be accepted.

Traditional investment analysis has relied on the concept of ROI, and because of the familiarity of this approach members of the financial community have tended to favour the other method of adjusting for the time value of money known as the Internal Rate of Return. The objective here is to solve for the rate of interest r at which the present value of future cash inflows is equal to the present value of cash outflows. This is described in the equation:

$$IRR = \sum_{t=0}^{n} A_t \div (1 + r)^t \qquad (9:3)$$

To solve for r in 9:3 requires an iterative approach to establish at what point IRR equals zero. The method can be demonstrated using the freezer truck financial data. It has already been established that a 20% interest rate will generate a positive return. The next step is to examine the IRR for a higher interest rate to determine whether the cash inflow is still greater than £36 000. If r is set at 22%, the following present values result:

Year	*Cash flow(£)*	*Discount factor*	*Present value of cash flow(£)*
1	11 200	0.820	9 184
2	12 500	0.672	8 400
3	16 000	0.551	8 816
4	14 000	0.451	6 314
5	8 000	0.355	2 840
-Present value			35 554

As this present value is less than the initial investment, it can be concluded that the value for r at which IRR equals zero is greater than 20% but less than 22%. A repetition of the iteration with r at 21% yields a present value for cash inflows of £36 455, indicating that IRR equals zero at a point somewhere between 21% and 22% interest rate. The precise rate is at a value of r equal to 21% + (£455 ÷ £901) or when r equals 21.51%.

An analyst's choice of method will often be based upon personal preference, although for a number of reasons the net present value approach is theoretically superior to internal rate of return. The NPV approach will therefore be used to analyse the shrimp farm investment example. As aquaculture is a high risk proposition, it must be expected that the discount factor would be set at a rate of 30%. Using equation 9:2 and discount factors from a table yields the following:

Year	*Cash flow(£)*	*Discount factor*	*Present value(£)*
0	-581 587	1.000	-581 587
1	44 081	0.769	33 898.3
2	162 629	0.592	96 276.4
3	1 138 403	0.455	517 973.4
4	1 138 403	0.350	398 441.1
5	1 138 403	0.269	306 230.4
6	1 138 403	0.207	235 649.4
7	1 138 403	0.159	181 006.1
8	1 138 403	0.123	140 023.6
9	1 138 403	0.094	107 009.9
10	1 138 403	0.073	83 103.4
Net present value			+1 518 025

On the basis of the positive net present value at the selected rate of interest of 30%, the shrimp farm can be judged a financially viable investment proposition.

FINANCIAL PERFORMANCE APPRAISAL

The outcome of investment decisions are later reflected in the profit and loss account and the balance sheet. To interpret fully the information summarised by these accounts, further analysis is required. The standard form of appraisal is that of 'ratio analysis', the results from which can be compared with the same ratios for the previous year. This can be illustrated by application to a slightly disguised annual report recently issued by a North American fish company. The elements of the ratio analysis are as follows:

'Operating efficiency' describes the profits as a ratio to sales and capital by using the formulae (profit before tax x 100) ÷ sales and (profit before tax x 100)

÷ long term capital. In 1980 the data for these two ratios were in $'000, (2 484 x 100) ÷ 274 447 and (2 484 x 100) ÷ 103 820, which yields 0.9% and 2.39% respectively. This result indicates the company has performed poorly, making less than 1% profit on sales and an ROI of less than 3%. Further insight can be gained by analysing the previous year's performance where the two ratios were (8 103 x 100) ÷ 268 057 and (8 103 x 100) ÷ 85 594 which yields 3.02% and 9.47%.

From this comparison some interesting, and in this case rather worrying, conclusions can be drawn. Firstly the decline in profits as a percentage of sales indicates a problem of costs rising at a rate faster than prices. Secondly in the current year, return on capital has fallen not just due to lower profits but also because the level of investment has risen from $85 594m to $103 820m. Company accounts are closely examined by individuals outside the company such as creditors, banks and shareholders. These results would be very obvious under such scrutiny, and the comments of the directors in the annual report would undoubtedly be examined very closely by these outside parties. In this example, it would appear that only a small component of the increase in investment was related to the injection of funds to underpin current sales, since $18 164 was invested in trawler and processing facility construction, none of which would be completed until 1981. Thus readers of the annual report can possibly be reassured that the 1980 performance downturn was of a short-term nature in view of the directors' comments that revenue was affected by 'two strikes in Newfoundland, but we consider our relationships with the workforce and our fishermen to be at a high level...'.

Liquidity is the term used to describe a company's ability to convert assets into cash in order to meet current liabilities. The only assets which would normally meet this criterion are those such as bank deposits, short-term investments and inventory. There are a number of ratios which can be used to assess various aspects of the liquidity issue. The 'working capital' ratio is calculated by dividing current assets by current liabilities. For the Canadian company the ratio is equal to in $'000, 106 522 ÷ 93 307 for the current year and 102 028 ÷ 81 194 for the prior year. The respective ratio values of 1.14 and 1.25 indicate that the company is moving towards a situation of possibly being unable to meet creditor demands unless profitability begins to improve.

Some financial analysts argue that it is unusual for inventory to be realised for cash unless a company is willing to sell finished goods below cost. An alternative approach which recognises this viewpoint is to calculate an 'acid test' ratio by excluding inventory from the current assets. In the fishing industry, finished goods often represent over 60% of current assets and companies are usually able to liquidate goods on-hand at cost value. Therefore most financial analysts in the industry feel there is little benefit from placing too great an emphasis on the results from an acid test calculation. This is fortunate in the Canadian company example because withdrawal of finished goods from the figures would in the current year reduce current assets to $39 876m. The resultant acid test would yield a result showing that assets are exceeded by 2.33 times current liabilities.

If inventory liquidity is of concern, further understanding of the situation can be obtained by calculating the ratio for 'average stock turn' using the formula (inventory on hand at year end ÷ cost of goods sold) x 12. For the Canadian company this is equal to ($66 646m ÷ $238 363) x 12 which represents a stock turn of 3.3 months. Most financial analysts would consider this result as an acceptably short period to conclude the company would be able to liquidate inventory in time to meet creditor demands. Where some question remains over this issue, the prior year stock turn can be analysed: if this is significantly lower, this would mean that the company is moving towards a less secure position to meet immediate demands for cash. Another indicator of the ability to meet cash demands is the average period of credit taken by the company which can be calculated from the formula (accounts payable at year end ÷ cost of goods sold) x 12. For the Canadian company the average period for credit was 1.0 months in the current year versus 1.2 months in the prior year. This result shows there has been no lengthening of the time taken to meet creditor demands; or to phrase this in another way, the company is not using their suppliers' money to solve their current cash flow and profit problems.

Most countries have legislation requiring companies to publish accounting records on a regular basis. Any company can therefore use this information as the basis of ratio analysis of their competitors in order to discover whether they are exhibiting a similar or differing financial performance. Should a competitor be achieving a higher ratio of profits to sales or investment, then further study of this company may reveal how to improve one's own operation. Conversely, competitors who are less successful may be considered a reduced business threat which may permit the reduction in promotional expenditure in the future. However, when using results from ratio tests on competitors, the analyst must be aware that variations in accounting practice can invalidate the comparison. If for example a competitor has decided to change the rate of depreciation on fixed assets or the method by which inventory is valued, this would dramatically alter the ratio of profit return against investment. Thus prior to carrying out comparative assessments, it is recommended that the annual report of the competitor is reviewed very carefully in order to gain a full understanding of the accounting methods being used.

8
Marketing and market systems

THE MARKETING FUNCTION

In the fishing industry, the widely accepted image of marketing is that it is simply responsible for promoting the product in order to increase sales. This very narrow perspective has been the cause of many of the industry's recent problems.

Since World War II, many other industries have realised that marketing has a much wider function than that of merely sales promotion. The first and most important role of marketing is to identify the product needs of the customers who make up the market sector in which the company operates. The second phase is to ensure that the company products offer the attributes most capable of satisfying need. Only at this point in the marketing process can the company begin to use the variables of price, promotion and distribution to generate the required level of sales.

Regrettably within the fishing industry there have been numerous examples of companies who failed to appreciate the full scope of the marketing management function and who invested in increased capacity without assessing the real nature of market need. In South America during the sixties, the whiting fishery was expanded on the assumption that the species would provide an adequate substitute for cod in the fish blocks used by US manufacturers of breaded portions and sticks. The nature of need in the fish block market is for a bland, white flesh, and processors who tried whiting blocks from South America soon found that the heavy fat layer in the subcutaneous flesh of the species led to rancidity problems in breaded products. If more careful attention had been paid to market need prior to the expansion of the fishery, the high level of rejection of these whiting blocks could have been avoided. Instead the fishery faced some years of major financial problems until in the early seventies the process of 'deep skinning' the fillets before freezing was introduced.

Failure to evaluate market need has also been the cause of some business problems in aquaculture. There is, for example, only a small market for many premium priced fish, and even with heavy expenditure on promotions to increase user awareness, the total market demand will always remain relatively low. Yet despite this fact, a large number of companies have invested in the construction of trout farms during the last ten years, only recognising with hindsight that production capacity is well in excess of market demand. This has led to fierce price competition, financial losses and the closure of some farms. Most of these events could have been avoided by paying closer attention to the nature of market need prior to the decision to build culture facilities.

MARKET RESEARCH

Evaluation of need and the development of appropriate products requires the acquisition of information through the use of market research. It must be stressed that market research will never provide the answer for a business problem, but will merely offer a manager some additional information with which to reduce the risk of making the wrong decision. All research is expensive; thus the objective in market research must be to only generate information which has a value greater than the cost incurred in its acquisition. There is little benefit in spending £100 000 on research if the results are to be used to reach a decision on a project capable of only generating £95 000 profit.

The lowest cost source of information is that which already exists, having been collected earlier for some other purpose. These data, known as 'secondary information', should be the first source examined in the research process. Within a company such information may be available from accounting records or from files created by the marketing department. Outside the company, there are other sources including Government statistics, trade association materials and articles published in magazines or newspapers which can be used to gain further insight on the issues being studied. In the event that the secondary information is inadequate, the company may decide to carry out specially designed studies to generate new (or 'primary') information. The standard methods for this include observation, experimentation and surveys. All three have their limitations which can result in errors, thereby reducing the validity of the results. Consequently, the choice of technique and the management of the error minimisation process is a specialist task which will usually require the services of a market research company.

Observation is used to study aspects of behaviour which cannot be analysed through surveys or by experiment. Consumer reaction to the presence of an assistant at a fresh fish section of a supermarket who can advise on product preparation and de-bone or scale fish may be the type of issue that can be assessed using observation. The technique is time consuming and the results only qualitative, consequently the method has only limited application in commercial market research.

It is often difficult to measure the influence of marketing such as price and promotion because of variations encountered in the market environment and the interactions which may exist between these variables. Experiments provide a mechanism whereby the influence of a single variable can be measured by comparing the results of a test situation against some form of control. For example a company may be interested in the effect on total sales of adding a second pack size of fillets to their product range. The control would be the sales volume in a group of stores which only stock the existing pack size. In the experimental stores the new pack size is added and the sales compared with the rate in the control stores. Although experiments appear to be an excellent research concept, in fact there are numerous potential causes of error which restricts their application to only certain types of marketing problems.

The commonest form of market research is that of surveys in which individuals respond to a list of questions or statements. The format of the survey instrument or 'questionnaire' will greatly influence the nature and depth of the response. Open ended statements (eg 'Why do you like crabmeat?') would allow the respondent to answer in their own words, possibly in great depth. However, this can result in the generation of large quantities of data which could present problems at the survey analysis stage. The analysis task can be reduced by presenting the respondents with a range of defined replies from which to choose. The most restricted form is that of dichotomous choice (eg 'Do you like crabmeat? Please answer yes or no.'), but if more information is needed a set of multiple choice responses would be used (eg 'Which of the following phrases explain why you like crabmeat. Flavour: Texture: Colour: Reasonable price: Ease of preparation: Can be served hot or cold: Will be eaten by the whole family.'). Restricted response provides a much greater degree of control and in large surveys is used more often than an open ended question format.

The lowest cost per response is achieved by using a mail survey, but this can only be done if the questions are sufficiently straightforward to be understood by an unsupervised respondent. There is always a time lag between mailing and return of the survey and another drawback is that some recipients may not respond at all. This lack of response may bias the results. A more rapid method of surveying in economically developed countries is through telephone interviewing. Most people are unwilling to spend long periods on the telephone and therefore the survey must be short, thereby limiting the quantity of data which can be collected. For very complex surveys, or those requiring some flexibility in the questioning process, it is necessary to use personal interviews. For consumer markets, interviews can be carried out by people in the street or calling at their home, and in industrial markets, by contacting respondents at their place of business. Personal interviewing can only be carried out by trained personnel thereby causing this to be the most expensive form of survey. However, the increased amount of information that can be obtained will more than compensate for the higher costs compared to postal or telephone surveys.

A fundamental assumption in market research is that data collected from a small number of people will provide results which are representative of the views of the entire market. It is critical that the selection of respondents, or 'sample', truly reflects the market as a whole. This can be achieved by the use of a probability sample in which every member of the population has an equal chance of being selected by the sample procedure. Unless some form of random sampling is used, many of the statistical tests to evaluate the validity of the conclusions reached cannot be applied. In certain circumstances, the random sample approach would be prohibited by high cost, and in this case a more qualitative result can be obtained by using a lower cost non-probability sample where respondents are selected on an arbitrary basis. (eg In a survey on brand preference to canned tuna, stopping a number of people in a supermarket mall, and accepting that this single location will decide the socio-economic profile of the survey.)

A recent example of the application of market research is provided by the University of Stirling study of the UK trout market in which a consumer survey of 952 housewives was used to gain insight on consumption patterns and attitudes to trout. The survey was executed using personal interviews and respondents were questioned using multiple choice and dichotomous questions. The results permitted the researchers to reach conclusions on issues such as:

(1) Trout consumption is biased towards the upper social groups, with lower social groups perceiving trout as being suitable only for special occasions.
(2) The most important reason for non-usage was high price.
(3) Regular trout users also tend to be heavy consumers of other fish products.
(4) There are no strong patterns of variation in consumption in different areas of the UK.

From these results when linked with secondary information, specific conclusions were reached on the need to increase product awareness and to define more closely the most appropriate socio-economic groups towards whom any future promotional campaigns should be aimed.

MARKET SYSTEMS

No company - unless it has achieved a monopoly - can carry out the marketing function without prior understanding of the market system of which the company is a part. At the industry level, the system is composed of five key elements:

(1) The producer company and the competitors supplying the market.
(2) The distribution channel linking the producer to the final customer. The channel may be simple, comprising only a few organisations, or very complex with numerous intermediaries between the producer and final consumer.
(3) The population which consumes the output from the fishery.

Surrounding the industry market system is a larger system known as the 'macroenvironment'. This latter system is composed of variables such as technological change, economic trends and Government legislation, all of which can influence sales volume in the industry system. The form of influence tends to be less defined than changes within the industry system and therefore will receive less consideration by the marketer in the planning and performance evaluation process. The hazards of this limited view have been recently demonstrated by processors in North America who ignored the signs of economic downturn at the end of the seventies and continued with plans for market expansion through the launch of numerous new products. Sales revenue during the current recession have been reduced by consumer buying behaviour modified in the face of rising unemployment. This situation has left some companies in a very weakened financial position.

The use of a market system model in marketing planning can be illustrated by examining an agro-industry company in southwest England who acquired some marginal land which had a river flowing through it. The Government grant situation and the company tax liability profile made it advisable to reinvest funds in the development of the acquisition. Due to the nature of the property, it was decided to construct a trout farm with an output capacity of approximately 50 000 lbs per annum. As part of the feasibility study, the company collected information from published sources and through trade interviews which provided them with some insight into the market system for trout in the UK (Figure 8.1). The existing producers were composed of a small number of very large farms with output capability of over 2 000 000 lbs per annum. However, the majority of farms in the UK are much smaller with a production capability similar to the proposed facility. The distribution channels for trout are both direct to consumers through 'farm gate' sales, and also through multilayered systems involving wholesalers and end user outlets. Consumption is split almost equally between private households and the catering sector.

The key variable in the system is the cost of production. This is because the large farms enjoy an economy of scale which allows them to raise trout at a cost significantly lower than the smaller farms. In addition, the dominant companies in the UK distribution channels prefer purchasing from large farms, not only because prices tend to be lower but also because reliance on one or two supply sources for their total needs can reduce the logistics of delivery and storage. On the basis of this information, the company decided to market their trout on a farm gate sales basis and embarked on further research to assess the size of the poten-

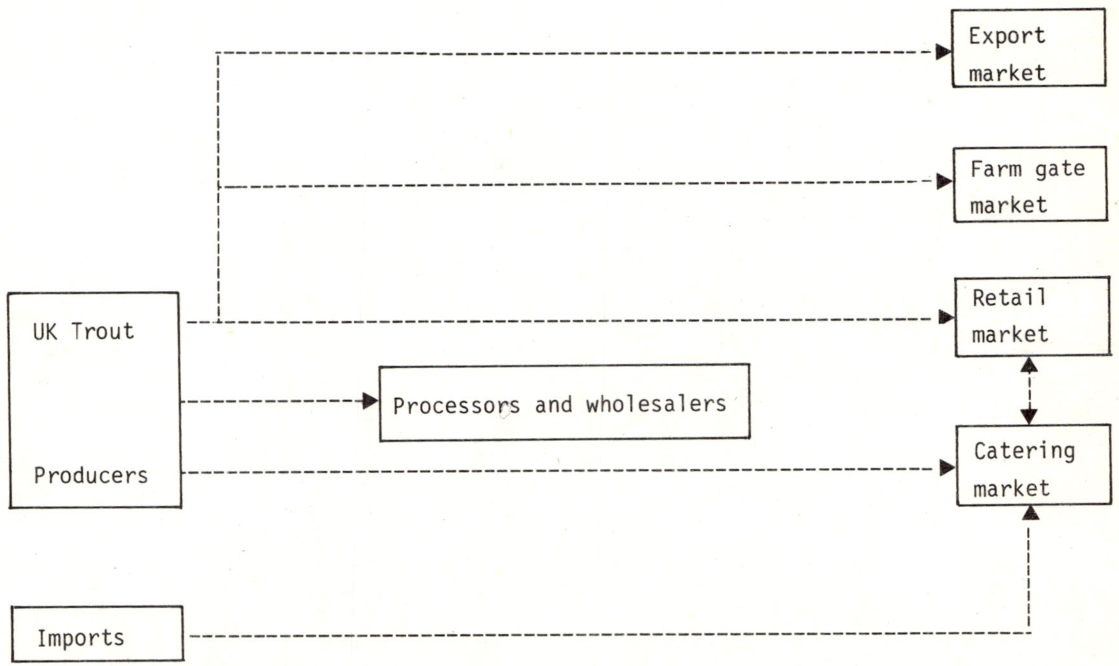

Fig 8.1 The market system for farm raised trout in the UK

tial buying population within a twenty mile radius of the site. This study also encompassed an evaluation of other farms in the area which might represent a source of competition. It was concluded that although a reasonable market potential existed, two other farms in the area had already gained a dominant market position. The company therefore decided that in order to succeed, it would be necessary to diversify into other trout-based products (eg pate and smoked fillets) and to expand the site to include a fishing lake and recreation area to attract incremental revenue from tourist visitors. Given the small output of the farm relative to the total industry, it was felt that little concern should be given to the macroenvironment variables and their possible influence on the market system during the planning process.

A common error in the use of market system models is that once a business is established, insufficient attention is paid to the implications of changes in the system on future performance. Hence it is recommended that companies include a system review as a standard component in the annual marketing planning activities. Issues which should be examined include whether user behaviour is varying from accepted patterns (eg a move from meat to fish in reaction to consumer concern about healthy diet), whether any of the distribution channel elements are losing or gaining dominance (eg supermarkets moving into fresh fish by opening in-store deli counters), and if competition appears to be revising their marketing strategies (eg competitors with only filleting operations who may decide to add capability to process shellfish).

The marketing department holds a unique position in a company in that its personnel are in constant contact with the business environment during execution of tasks. This means that the marketing group must be orientated to identifying new opportunities or problems, not only those which affect their role but also indicators of change which could have impact on other departments. For example, it is usually the sales staff who are first aware of new competition when customers inform them of a better price quote. This information when passed onto the procurement group could cause them to review ways of acquiring lower prices. A number of shrimp companies in the late seventies ignored developments in aquaculture and were proceeding with plans to expand trawler fleets. Their sales forces began to encounter price competition which upon investigation was caused by aquaculture groups from South America entering the market for shell-on shrimp. It was only at this stage that the established companies in the industry began to seriously consider that aquaculture would change the supply situation. These companies are now attempting to invest in shrimp aquaculture. It has to be expected that acquisition of the necessary expertise at such a late stage will at best be expensive, and at worst virtually impossible, because the farming groups may have developed sufficient dominance in the market that they can resist any entry attempts from the traditional sector of the shrimp industry.

9
Product marketing

PRODUCT DIFFERENTIATION

Historically the standard approach to marketing was to supply a uniform product on the assumption that this would maximise profits. The ultimate point in such a philosophy would be for all to sell the same product which would provide no opportunity for the customer to distinguish between the various offerings. A major component in modern marketing management is the concept that the company must endeavour to establish a unique position in the market for its products, thereby creating a recognisable identity for the potential customer.

There is significant resistance to this latter concept in the fishing industry because it is argued that the uniformity of output from a fishery or from aquaculture vitiates the possibility of product differentiation. It appears not to be understood that all elements of the product - physical characteristics, pricing, promotion, distribution - can be combined to achieve a product identity. An illustration of this is Icelandic cod which is only perceived as a premium quality product because of the totality of the market presentation (ie fillet cuts, grading, packaging, quality control and promotional campaign). Even when shown examples of this type, there is still rejection of the concept and it will undoubtedly remain a point of contentious debate in the future. It would seem more than a coincidence, however, that the more successful companies in the industry are those who have established a distinctive image in the market. Whereas commodity orientated firms who base their business strategy simply on price competition do seem to get into financial difficulty with a distinct regularity.

The first step in the product positioning process is to evaluate which product characteristics are important to the potential customer. This may be achieved through a market survey or, where customers are unable to articulate their basis of need, more sophisticated techniques such as non-metric multidimensional scaling. The research will usually reveal that a market is composed of subsegments of different needs known as product spaces. For example in the catering sector for scallops, there are numerous product spaces reflecting differing size, colour and texture needs depending upon whether the meat is to be served unmodified, used in a seafood platter, coated with batter or an ingredient for a stuffing.

Having defined the product space, the next step is to determine which of these is the most appropriate for the company's product. It is preferable to enter a product space where there is no competition. Where, as is often the case, it is impossible to achieve a product space monopoly, then further research and possibly product modification may be necessary to identify the basis for a superiority claim. For unprocessed fish and seafoods, there is little opportunity to modify physical characteristics, and therefore it is often necessary to examine other areas such as quality control, grading, packaging or distribution to gain an 'edge' over competition. The same solution may apply to processed goods, but in this case the manufacturing process offers numerous opportunities to modify the product's flavour, texture, size or, by the substitution of ingredients, reduce costs. For this reason many larger companies in the industry endeavour to maximise the sales of processed products because they can be more certain of achieving product differentiation.

Having established a unique positioning, the product identity can be reinforced by the use of a brand name. The brand can be applied to a single product or across the whole range of company goods. The former approach tends to be used in industries where the profit margin is high, because this provides the funds to support the large promotional budgets required to market products individually (eg toiletries and detergents). The profit margin is usually much lower in the fishing industry and thus the tendency is to market a range of products under a family brand name. In this situation, the promotional activity on one item will create greater brand awareness for the company's other products which are otherwise unsupported because of insufficient financial resources.

THE PRODUCT LIFE CYCLE

During the life of a product, sales volume can be divided into four distinct phases, namely:

(1) The 'introduction' phase when a limited number of customers are trying the product for the first time and total sales are low.
(2) The 'growth' phase during which initial purchasers adopt the product as a regular component of their overall consumption and the remaining potential customers are trying the product for the first time.
(3) The 'maturity' phase when the product has attracted a group of regular users and sales reach a plateau.
(4) The 'decline' phase when the product's former users switch to other goods and eventually the sales trend to zero.

Although it is virtually impossible to predict the length of time over which these various phases will occur, the marketer has to recognise the strategic implications of the life cycle process and its potential influence on product positioning. During the introduction phase, the marketing objective is to maximise the trial rate and this can be achieved by using the broadest possible product positioning in order to attract a large proportion of the buying population. Trial will only occur following the purchaser's awareness of the product's existence and therefore funds will have to be expended on promoting the availability of the new item. It is often the case that this promotional expenditure will exceed sales revenue which causes a negative profit during this phase of the life cycle. Growth occurs as more people acquire the repeat purchase habit. In order to sustain this activity, continued promotional expenditure will be necessary, but this will be accompanied by increased sales and a positive profit position will be achieved. The company's success will attract competition and so the marketer will begin to concentrate on product differentiation by greater emphasis on a more specific benefit positioning than that used in the introduction phase. This approach has been seen recently during the launch of frozen seafood entrees in certain Western countries. Initially the manufacturers promoted ease of preparation and increased variety for menus. As the sector began to grow, emphasis turned towards specific positionings such as calorie controlled diets or gourmet cuisine.

At maturity, no incremental sales can be generated by attracting new users to the market. The only source of sales is that of existing users, and the promotional objective will be to protect the product from loss of customers to competitive brands. Thus maturity is the phase of the life cycle during which greatest emphasis is placed on consolidating the product positioning. The majority of market categories in the fishing industry are in the maturity phase and, without exception, it is only those companies perceived as offering a unique product benefit who remain successful over the long-term. Complacency will inevitably be followed by failure, so the responsibility of the marketer is to strive to improve the product's perceived benefit. As has been mentioned earlier, product improvements are technically easier for processed goods as the manufacturing process permits the introduction of various modifications. A number of processors used this approach in the US in the late seventies by consolidating their market position through the introduction of new types of coated products (eg battered fish portions which could be cooked in the oven instead of the more traditional deep fried method; flavoured batters; textured breadcrumbs to enhance crispness).

Although unprocessed products can sometimes be physically improved (eg the introduction of boneless trout by Idaho processors), it is more usual that market position consolidation will occur through changes to other components of the marketing mix. These can include product presentation (eg a move from block frozen to individually frozen fillets), packaging (eg inert gas pack fresh fish to extend shelf life), and moves to new distribution systems (eg catfish farmers opening restaurants instead of selling output to wholesalers).

There are major economies of scale associated with multi-product marketing and the added advantage that through the wider range of products, customers' recognition of the company brand name is more easily achieved. Hence another defensive measure at maturity is to expand the product line by the addition of items closely related to the leading company product. In this way it may be possible to increase sales by moving into an additional product space (eg a producer of smoked salmon who adds a range of smoked shellfish as a product line extension). Although product line expansion and sustained promotional activity can extend the length of the maturity phase, ultimately all products will move into decline. Within the fishing industry, the probability of decline is increased by excessive exploitation causing landings to fall, when in turn customers switch to alternative, lower priced species. At the point of decline, the greatest hazard is that the company ignores the indications of an impending drop in sales and continues to invest in promotion to halt the downturn in consumption. The result of such action will be to erode profitability and to focus management's attention on existing products at a time when they would be more productively employed on new product development.

A characteristic of the industry is that in many sectors, companies sell a diverse range of species. When one of these species enters the decline phase, it is not removed from the product line because of the possible discontent which could be aroused amongst intermediaries in the distribution channel. This is an unfortunate philosophy because not only is the company making less profit than if efforts were diverted to other products, but intermediaries would benefit from using their warehouse and shelf space to stock another product which is either in a growth or maturity phase of its life cycle. This fact is now being recognised by the more innovative processors, and their relationship with intermediaries has been strengthened by their voluntary withdrawal of weaker products from the market. These companies may experience some reduction in sales over the short term, but this is compensated by increased co-operation and buying loyalty among members of the distribution channels.

NEW PRODUCT DEVELOPMENT

It is increasingly evident that the time between introduction and decline for new products in the food industry is shortening. The implication of this is that even in the most successful companies, there should be at least one new product under development to eventually replace every one at the maturity and growth stage.

The main source of new product ideas are changes in customer behaviour (eg the increase in working women, which expanded the need for convenience foods), and the exploitation of new technologies (eg extrusion machines which made possible the manufacture of imitation breaded shrimp products). In the fishing industry, another possible idea source is to forecast which species may decline in availability due to overexploitation of natural stocks. It may then be possible to locate an alternative raw material which can be marketed (eg the Alaskan pollack fishery replacing high priced cod blocks), or to develop new processing methods which could extend raw material supplies (eg the use of soya as a substitute protein in certain processed fish products).

The risk of failure in new product development is high, with recent estimates indicating a failure rate of over 80% of products launched. Certain guidelines should therefore be defined by the management for the development process. Lack of knowledge in the area of interest can increase the risk of failure, and in recognition of this fact it is advisable to concentrate on markets and product types where the company has previous operating experience. For example, if a company raises tilapia for sale into retail markets, it should examine opportunities to raise other finfish or develop a processed tilapia-based product rather than invest in a closed system shrimp farm linked to a chain of seafood restaurants. Only when the company is satisfied that all further opportunities related to current business activities have been exploited, should the search for new ideas be extended to areas outside current experience.

The most expensive point for a new product to fail is after its launch. Emphasis should therefore be placed on control procedures to identify probable failure before this point in the development programme is reached. The first point of control should be at the idea stage when the new concept is reviewed relative to such issues as expected sales, the nature of existing or potential competition, compatibility with current processing/marketing activities, effect on other company products, and the degree of technical complexity associated with manufacturing the item. Only if the idea ranks highly in this type of screening should the project continue. Product design will require precise definition of potential user needs, a factor which should be evaluated through a market research study, not simply based on the judgement of company executives. This research will probably include group discussions to assess reaction to the product concept, to be followed after prototype development by placement studies to assess if the new item meets the expectations created by the product concept.

Where the product is to be sold to high volume outlets in the food service sector (eg a chain of hotels), the earlier research will provide sufficient information on which to decide whether or not to launch. Consumer markets are more diffuse, however, and other factors including the type of promotional campaign and reaction of competitors will have to be taken into account when evaluating the likelihood of success. A technique known as 'test marketing' is often used in which the product is placed in a restricted geographic area and sales performance monitored. If the test is executed in a number of markets, this can also provide an opportunity to evaluate alternative promotional programmes. The duration of the test is critical. It should run for a sufficient length of time to measure the number of people who try the product, the proportion of these who repurchase on a regular basis (the 'repeat rate'), and the average period between each purchase. If the trial and repeat rates are high, the decision will be to launch the product immediately. A test market result of high trial/low repeat would indicate that the new product is failing to meet customer expectations. It may be possible to overcome this problem by a product re-design. Where the trial rate is low but repeat rate high, the solution is to improve the promotional campaign. Occasionally a product will exhibit both a low trial and repeat rate, indicating a fundamental problem in the product concept. Except in the rarest of circumstances, the best decision in this case is to accept that the product is a failure and halt all further development plans.

The risk in executing a test market is its possible identification by competition, thereby providing them with a clear understanding of the company's plans. Usually this risk is far outweighed by the greater costs of launching a failure; thus test marketing is considered indispensable by the major companies in most industries. Nevertheless to test without revealing any information is the ideal of market researchers, and in recent years a number of companies including some in the fish industry, have successfully experimented with computer simulations of product viability in place of a test market.

The majority of companies in the fishing industry have yet to recognise the need for the effective management of the new product development process. Many rely on the continued success of a single existing product, and when sales enter the decline phase these companies lack the financial resources or the time needed to develop an alternative revenue source. The risk is possibly greatest at the catching and primary processing level of the industry where an unexpected reduction in landings is usually followed by widespread company closures (eg the pre-war pilchard fishery in

Cornwall; the anchovy fishery in South America; the herring fishery in the North Sea). Despite these examples the aquaculture sector is repeating this error, especially those groups exploiting a single species system where there is a reasonably high risk of disease through contact with 'wild populations'.

10
Managing demand

PRICE

Average market price is controlled by the economic laws of supply and demand. On the demand side, as price decreases the unit sales in most cases will increase. In cases where supply is flexible, then more units of product will be made available as prices increase. Market price will be the equilibrium point at which the supply and demand curves intersect.

For the marketer, price is an important variable used to influence demand for company output. Also, unless the price at which the goods are sold exceeds cost, the company will not generate a profit. Price is also an important variable by which potential users evaluate the validity of a company's product positioning, a factor often ignored in the fishing industry. In a situation where a company has decided upon a positioning of superior quality, it is necessary to reinforce the validity of this characteristic by offering the goods at a comparatively high price. Equality of pricing with competitive items will usually only be effective if the company's goods are of a similar quality. Conversely, if the product positioning is that of below average quality in order to maximise unit sales volumes, this positioning should be reinforced by lower than average prices.

A common error in the industry is the assumption that a company can build a secure dominant share of the total market by pricing at a level significantly below competition. It is frequently the case that the company concerned has achieved an economy in their operation which makes the reduced price viable (eg a new fishery where landings per unit of trawler effort are higher than that in a more mature fishery; an aquaculture venture following a technical breakthrough in productivity per acre of ponds). However, the chosen pricing strategy ignores the fact that the market will associate low price with low quality. Furthermore, it is almost inevitable that the company will eventually face either rising costs or a market entry by a competitor who has achieved another form of operating cost economy, at which point the original company will be unable to compete because customers will switch to the competitor's new lower priced item. It is advisable, therefore, to approach the decision on pricing with caution in order to ensure compatibility with the required quality positioning. If a genuine operating economy is achieved, it is more effective to price at a level equal to existing competition and to reinvest the company's higher gross profit in promotional activity designed to build user loyalty.

The commonest form of price setting in the fish industry uses the formula Price = Product cost + Required profit. This is known as 'cost plus' pricing. If a scallop processor using this method had estimated his cost elements were meat £2.50/lb, processing £0.40/lb and distribution £0.10/lb, and his profit target was £0.50/lb, then the quoted price for his product would be £3.50/lb. The major drawback in cost plus pricing is that it can only be effective if all the suppliers in a market have similar operating costs and share the same requirement for profit. In practice, individual profit requirements will vary, as will operating costs (eg a company with a new processing plant would face much higher depreciation charges than one which has operated their facility for over ten years and has a fully depreciated asset). Furthermore, in the final analysis it is the market demand curve and not the expectations of suppliers which set prices. It is therefore more sensible to use a pricing policy which reflects market forces. This approach, known as 'demand pricing', recognises that there is a direct relationship between price level and quantity sold. A company decides on the quality/price appropriate for market conditions; any price changes at a later date will be in response to cost variations or competitive activity. Thus if the current price range in the market for breaded shrimp is £2.50 - £3.50/lb, then a shrimp processor considering the launch of a premium quality brand, using a demand pricing policy, would set price somewhere in the region of £3.40 - £3.50/lb for the new item.

Demand pricing requires a full understanding of market responses to varying price levels, and this can prove difficult for smaller companies with limited marketing expertise. For these organisations, the most appropriate policy is that of 'price follower' in which price is set at a defined level relative to the dominant company in the market (eg equal to the brand leader or 5% above or below the brand leader). Once this relationship is defined, then all the company has to do in the future is revise prices in direct proportion to any change by the dominant brand.

As so many companies in the industry rely on price to manage sales target achievement, the marketer will frequently have to respond to price changes by competitors. The standard response to price decreases is to match the new price or do nothing. Whereas in fact, unless the decrease is merely a short term tactical move to reduce inventories, the marketer must commence a more thorough analysis of the situation and respond in a form appropriate to the magnitude of the price decrease. When the price reduction is small (eg less than 3%), it is likely to have little impact on current customers and can be ignored. A change of 3-6% is relatively unimportant and an effective response to blunt the impact of the move is to introduce a short-term promotional programme (eg a 10% merchandising allowance for 30 days; distribute money off coupons good on next purchase). Where a competitive price reduction is between 6-10%, a more significant reaction is needed. Usually this should take the form of a major increase in promotional programmes such as advertising in the consumer markets and expansion of sales force market coverage in the food service sector. In the rarer situations where a competitor reduces prices by more than 10%, the marketer will need to carefully re-examine the viability of the company's product to assess the practicality of a relaunch based upon a tangible product improvement. When this type of strategic redirection is not feasible, then consideration should be given to product discontinuation. This apparently drastic solution is necessary, because if the competition has indeed acquired a genuine operating economy which cannot be duplicated then their price advantage will permit them to rapidly take over the market. There is little point in attempting to halt their expansion; it is better to redirect company resources into another more profitable area of operations.

The fishing industry is composed of many small, independent operations all of which strive to obtain the highest possible price for their product. Each company's willingness to accept the price offered by an intermediary will depend upon their profit objective and financial resources. Understandably the intermediary will bid at a low price in the hope that the weaker members of the industry will accept.

This in turn may lead to a more general decline in prices. On the other hand, when supplies are scarce producers will attempt to bid up prices by withholding product from the market. Both types of activity create a volatility in the market which eventually can cause consumers to switch to protein foods which are more stable in price. Because of this risk, a growing number of primary producers are now members of marketing co-operatives. This type of organisation, as a major supply source in the market, is able to achieve price stability which is of benefit to its members and their customers. This form of price stability management, which originated at ports where many independent trawlers land catch, is now spreading to some sectors of the aquaculture industry (eg trout farms in the US; salmon farms in Scandinavia) where individually, each farm has only a small annual production capability.

PROMOTION

A popular misconception is that promotional programmes will sell any product. In fact, unless the product is capable of satisfying market need and is offered at an acceptable price, promotions will not influence potential purchasers. The real role of promotion, therefore, is to communicate to potential purchasers information about product benefits compared to competition, thereby increasing market awareness for the product. The most effective form of communication is a dialogue in which the participants can exchange information and respond to questions. The implication of this is that the best form of promotion is personal selling in which a company representative can discuss the product with the potential customer. The personal selling process is composed of the following elements:

- Identifying the potential customer.
- Comprehending the potential need of the customer.
- Developing an effective sales presentation.
- Creating a situation in which the sales presentation can be executed.
- Managing the objections from potential customers.
- Administering the purchase transaction and arranging delivery.

All of the above are both time consuming and expensive. Costs will be reduced if the customer is able to buy a large quantity of product and the distance between customers of this type is small, which will minimise sales staff travel time while making calls. These conditions prevail in the food service sector because end users such as hotels, restaurants and canteens all use large quantities of product and tend to be located within a reasonable distance of each other. Consequently personal selling is the most prevalent form of promotion in this sector of the industry.

This situation contrasts with consumer markets where the end users are private households spread across the country. This is a diffuse market structure and although personal selling could be effective in building brand awareness in this market, the costs incurred would greatly exceed the potential sales revenue. Under these circumstances the mass communication vehicle of advertising is the most appropriate and cost effective promotional mechanism.

The two factors influencing sales force structure are customer needs and cost. The low profit margin in the fishing industry means that companies favour the sales agent structure, whereby another organisation represents the company in a market area and is paid by a commission on sales. In this way the company only incurs expense in the event of a successful sale. A potential drawback to this approach, however, is that the company has less control than if its own employees executed the sales function.

Where the company has a relatively homogeneous customer base and a simple product range (eg a fillet processor), the usual solution is to organise the sales force on a geographic basis with each individual responsible for all the customers in their specified area. As the market becomes more complex (eg a catering market which includes restaurant and hotel chains demanding products to fit standardised menus, and centralised public sector canteen buying groups who need recommendations on product forms to suit fixed budget buying programmes), a need is created for specialists within the sales force who can manage these differing customer groups. Then the geographic organisational structure will tend to be replaced by one in which sales personnel are allocated by customer types.

In markets where the primary promotional vehicle is advertising, the role of the sales force is to manage the company's relationship with intermediaries such as the major retail chains and larger wholesalers. Although the sales force is responsible for generating orders from these intermediaries, the focus of their activity in consumer markets is to work with the trade on merchandising programmes, product flow logistics and in-store management of the product. This will ensure that the consumer will encounter the product in such a way as to reinforce the preference over competitive brands which was initially created by the advertising component of the promotional mix.

Newspaper or television advertising is an ephemeral form of communication. The advertising message must therefore focus on a single aspect of the product if the campaign is to succeed in building a high level of consumer awareness. The development of effective advertising campaigns is a specialised task which the company should delegate to an advertising agency. The agency will also be responsible for recommending the most appropriate media in which to execute the campaign. In economically developed countries, the medium of television can reach a larger proportion of the population than any other. In contrast, by using trade or consumer magazines, the advertiser can be very selective over which sub-groups of the population to whom the promotional message is directed. The choice of media in underdeveloped countries tends to be very different with radio or billboards used for mass communication. Depending on the literacy level, however, the medium of print provides a mechanism for reaching selected sub-group targets in the population.

In order to build awareness through advertising, the same message must be repeated frequently as, on average, an advertisement has to be seen at least three or four times before it has any significant impact. It is extremely expensive to execute this level of repetition across the majority of the population and hence major campaigns are beyond the capabilities of most companies in the fishing industry. Therefore many of the industry's promotional campaigns are, at best, a compromise. For individual companies the tendency is to restrict advertising to only a few geographically restricted markets, and for those firms with an extensive product range, to limit advertising to just one or two of their leading products. Many smaller companies cannot afford any advertising except possibly a campaign executed by their trade association or marketing co-operative. Although this solution does create some market awareness, product recognition will be of a generic nature and not for any single brand name or company.

In consumer markets it is almost impossible to build a stable business based on user loyalty without using advertising. Unprocessed products can rarely provide the profit margin necessary to fund advertising campaigns, and for this reason larger companies emphasise building sales for processed goods because these can generate the profits required to fund campaigns. This situation for unprocessed goods may change in the future if aquaculture groups use their savings from lower production costs for promotional spending and not as at present, merely use the cost advantage to undercut their competitors on price.

Personal selling or advertising are the only consistently successful methods through which to achieve high brand awareness and customer loyalty over the long-term. There are, however, other promotional tools which offer some limited advantages in managing promotional demand. A useful tactical weapon is to provide a temporary value incentive to subsidise the purchase for a limited period (eg a coupon good on next purchase, an in-store programme offering a special low price). This type of activity is extremely popular in the industry, but unfortunately what is not realised is that this type of promotion will only increase sales during the period of the offer without any benefit to long-term sales trends. Another form of promotion is public relations, where the company or a group of companies provide information to the media in the hope that articles will be published or mention given in television or radio programmes on some aspect of their activities. This news is communicated at no charge (cf the purchase of advertising space) and hence it must be realised that the details may be modified by the media or not used at all if an alternative event occurs which is of more interest to the general public. Despite this major drawback, public relations remains a popular component of the promotional mix in the industry, apparently being perceived as a low-cost form of advertising. It is probable, however, that in many instances the only real beneficiaries from these campaigns are the public relations firms which are retained to organise the events.

DISTRIBUTION

The move to an urban society which accompanied the industrial revolution caused many sources of food production to be left some distance from major centres of population. The process of moving goods from the site of production to the point of consumption is known as distribution. In theory it would be most advantageous for the producer if he were in sole control of the entire delivery system. In practice, however, the use of intermediaries such as wholesalers and retailers can offer significant operating economies which result in lower prices for everybody. These economies are based on the fact that an intermediary will reduce the number of transactions in the purchase process and will combine a range of different products into a form most favoured by the final purchaser (eg a supermarket which accepts truckloads of bulk packed fillets from different processors, repacks the fish into trays and displays the various species in a special fresh fish section in the store).

The fishing industry tends to blame intermediaries for many of the current problems, claiming that the 'middlemen' demand excessive profits and their mark-ups result in high prices which reduces customer demand. However, in almost every case where this complaint has been examined fully, not only were no excessive profits being made, it was found that the intermediaries were providing a distribution system at a lower cost than any alternative method, unless the final customer was prepared to come to the point of production to purchase (eg farm gate trout sales).

The rising cost of energy means that distribution represents a major expense component in the industry, sometimes up to 40% of total costs. This has led marketers to accept the responsibility for managing the distribution process with the objective of minimising costs - thereby keeping down the price of the product - through the selection of the most appropriate distribution system. In addition to the obvious costs of warehousing and transportation, the influence of time on product quality is a 'hidden' cost in the distribution process.

The highest quality product is that which is fresh. Where markets are not situated near to the point of production, the marketer either has to use fast, but expensive, methods of distribution such as air freight, or accept a reduction in quality and use some form of preservation, typically freezing, smoking or canning. Even with preserved products there are choices of distribution which include road, rail or water with systems involving the longest delivery time being offered at the lowest cost. Unfortunately the longer the delivery time, the greater is the risk that the customer will buy from another supplier who can deliver more rapidly. Thus in reaching an optimum decision on the best form of distribution for the company, the cost savings of selecting a slower delivery system will have to be weighed against the cost of lost sales to competition.

To an increasing extent, many distribution channels are becoming dominated by a small number of large intermediaries, supermarket chains or fast food restaurant groups. Although such companies are capable of purchasing the entire output of many fish companies and thereby represent a cost effective distribution path, these intermediaries may use their dominance of the market to demand heavy price subsidies from their suppliers. This means that a company's financial destiny is in the hands of a single intermediary, and in some sectors of the industry producers have chosen to create a wholesaler co-operative in order to retain control over the distribution channels. Occasionally the problem of intermediary dominance is solved by a producer ignoring the traditional channel and establishing an alternative pathway through which to sell their output (eg shellfish farmers who started their own group of oyster bars).

The marketer must never assume that the chosen distribution will always offer the most appropriate system in the future. A characteristic of all channels is that they change in response to trends in customer behaviour and adverse operating costs (eg the demise of high street fresh fish shops in Britain in the face of rising rents and local taxes; the expansion of fish and seafood restaurant chains in America in the mid-seventies as consumers became more aware of 'healthy eating'). Therefore the marketer must regularly evaluate the company's relationship with intermediaries and seek new opportunities to expand into other channels which have the potential to be a major source of sales in the future.

11
Production management

PLANT LOCATION

The primary processing sector in the fishing industry is usually located near to the point of landings. As a large proportion of catch is discarded during processing (eg the frames of fish after filleting; the shells of molluscs after shucking), this location means that the only transportation costs are those of delivering the finished product to the market. As many fisheries are to be found in underdeveloped countries or in areas of low economic activity, this type of site frequently offers the added advantage of low labour costs.

The non-fish components of highly processed products (eg breaded portions, seafood entrees) tend to be very bulky and expensive to transport. The more recent trend, therefore, has been to site facilities for these product forms nearer to the final market. For companies who export, the location of the production operation inside the final market may have the added benefit of avoiding the tariffs on processed goods shipped into the country from overseas.

Selection of the optimum site for aquaculture is more complex, taking into consideration variables such as the biological environment for the species cultured, availability of adequate supplies of clean water, resource conflict with other potential users of the site, and accessibility of the technical services associated with the management of the system. Furthermore, in certain circumstances the actual site may have influence on the type of culture technology to be used. In an underdeveloped country the combination of variables may lead to the selection of an extensive grow-out system, whereas in an economically advanced nation, the cost of land or labour may mean that a capital intensive, closed system would be the most cost effective.

PRODUCTION LAYOUT

In designing a new production layout, the first decision is that of the required plant capacity of the system. This will then influence the plans for handling raw materials, the size of working area for personnel and machinery, the inventory system for finished product, and the services associated with the production operation such as utilities, employee facilities, maintenance areas and access roads.

Whenever possible, when a new processing plant or aquaculture system is being planned, arrangements should be made to study existing industry layouts or, if this is not possible, to employ engineering design consultants who have experience of similar projects for other companies. It is frequently the case in aquaculture that no previous knowledge exists on the proposed system. Under these circumstances, it is advisable to build a pilot system from which information may be obtained that will more than recover its costs in terms of lessons learned when the full scale facility is constructed.

A common error in the fishing industry over the last twenty years has been the construction of processing facilities which significantly increase the output capability of a company, based upon the assumption that sales volume will continue to increase. In reality the pressure on fish stocks has pushed up the cost of raw materials and when reflected in finished goods prices, this has reduced total industry volume. There is little indication that this situation will change in the future and therefore companies would be wise to adopt a very conservative position on projected sales volume when planning new processing facilities. There is some evidence that aquaculture provides a major economy scale with the larger farms achieving the lowest production costs. This had led some companies to construct very big farms at the outset in order to maximise the cost benefits. In many cases, however, these new large farms have encountered biological stock management problems which were not encountered in their small pilot-size operation. It would seem advisable, therefore, to plan for a phased expansion programme in order to learn how to manage the greater output on a more gradual basis.

Most production management texts classify production systems into three main types: fixed position, process, and product layout systems. The fixed position layout is used where, because of its size, the product remains in a fixed location, and raw materials, sub-components and services are brought to the site. This situation only occurs in the trawler construction sector of the industry where the new ship remains at a fixed point in a construction yard until launching. A characteristic of the fishing industry is that many companies produce a large variety of goods in small quantities (eg a processor filleting various species of groundfish; a company manufacturing a range of breaded speciality shellfish items). This type of output can most effectively be handled using a process layout in which all the machines or processes of the same type are grouped together in one location. Thus in a plant producing frozen fillets, the cutting tables and skinning machines will be in one area, the grading and quality inspection in another, packaging and weighing in a third, and the freezing of the product prior to final storage in a fourth area. This type of layout offers flexibility in production volume, product mix and variance in product processing characteristics, because a problem in one area of the operation will not cause a total stoppage throughout the workforce until a solution is found.

Against the important benefits of production flexibility for the process layout approach must be weighed the costs associated with moving materials from one area of the plant to another. Where a company is producing large quantities of a standard product this 'interprocess cost' could be reduced by using a product layout approach in which the machines or processes are placed in the manufacturing sequence, thereby creating a production line. Although common in industries such as cars and electronics, product layouts are only of benefit to a minority of companies in the fishing industry who produce large quantities of each product. Examples of product layout can be found in the plants operated by the dominant companies in the frozen, coated fish or seafood sector in North America and Europe. These companies may have a production line which processes only fish fingers, another for breaded portions, and possibly others for additional high volume items. By designing a specialised product layout for a specific item, it has been possible to achieve throughput rates ten times greater than that attainable for the same item using a process layout. The major drawback, however, is that a single breakdown on a line (eg a transfer belt jamming between the fryer and freezer on a breaded portion line) can bring the whole production process to a standstill.

Whichever layout is adopted, a key objective must be to balance the workloads so that the same amount of effort is required at each work station. This will result in efficient overall performance and a satisfied work force. The balancing of effort across the plant will usually involve some form of work measurement technique to evaluate labour input at various stages in the production process.

PRODUCTION SCHEDULING

Fundamental to efficient production scheduling is an accurate forecast of demand. In a company which embraces the concept of strategic management, the responsibility of forecasting will lie with the marketing department which uses methods such as market trends, economic conditions, competitive activity and market research to develop the forecast. The recent advent of lower cost microcomputer systems has had a major impact on companies' abilities to produce very accurate sales volume predictions.

It is sometimes claimed by members of the fishing industry, that there is little point in sales forecasts because actual output is controlled by landings whose fluctuations cannot be predicted. The production managers in these companies then pack what they feel is best for current market conditions, only to find that they are left with the wrong product on inventory at the wrong time. A preferable approach where raw material shortages are expected because of signs of a downturn in landings, is for the company to form a production co-ordination committee composed of staff from both production and marketing to discuss the most appropriate product to pack. Decisions from this committee can also be communicated to the sales force who will then be in a much better position to inform customers of likely supply problems before they occur.

The other important issue in the production scheduling process is the company policy on finished goods stocks, the alternatives being to produce goods against anticipated orders or to approve production only upon receipt of an order. In the fishing industry, the chosen policy is heavily influenced by the fact that both raw material availability and customer demand exhibit seasonal variation: if a company was to wait for orders prior to scheduling production, raw materials may not be available at the time because landings have ceased and/or their processing capacity may not be sufficient to meet all the demands of customers in a given period. Consequently, the majority of the industry schedule production against anticipated demand by processing on a year-round basis and holding finished goods on inventory to service orders received. Where individual customers have specialised needs, for example a processor who produces a stuffed, breaded oyster using a formula developed for a specific restaurant chain, then the tendency is not to schedule production until after an order has been received, unless the customer is willing to commit to a contract which defines a regular delivery pattern without sudden cancellations or changes in order size.

The two, often conflicting, objectives of the production department are to satisfy customer needs by having products on inventory to meet any order size, and at the same time to minimise the costs of production. These costs can usually be divided into two components: the actual cost of manufacturing and the costs of storing the finished product. Typically the greater the quantity of an item produced during the production run, the lower the unit cost of production. This is because the plant is not losing time changing between different line set-ups, and because by concentrating on one product the workers become more efficient. However, the greater the output the higher the storage costs because the finished goods will remain on inventory longer before being sold. Thus the principle of cost minimisation is to balance output costs against storage charges. One approach is by defining an 'economic order quantity' or EOQ. Various techniques exist for calculating EOQ and the most common is based on the formula:

$$EOQ = \sqrt{(2 \times P \times D) \div S}$$

where

P = the cost of setting up production
D = estimated demand over a defined period
S = the cost of storage

Once the EOQ levels have been established, the production department can monitor on-hand finished goods and when these fall below the predetermined minimum, additional production can be scheduled.

Once the production department has established an outline of the expected processing schedules over a defined period, it must execute another responsibility, namely ensuring that raw materials are available when needed. This area of planning is complex because of (a) the lead times involved in acquiring materials from suppliers and (b) the need to achieve a balance between the savings of buying in large quantities and the costs of holding materials at the plant for a significant period before their use in processing. In companies in the processed goods sector, the procurement function is further complicated by the use of fish and non-fish ingredients for the products. The usual solution is to have a manager responsible for co-ordinating purchases from the fisheries which supply the company and another individual responsible for the other items such as oils, breadings and packaging.

A common theme in all the roles of the production department is that of achieving a balanced processing schedule to avoid peaks and valleys in output. In many countries, for example, customer demand peaks during Lent. To keep production running smoothly, processors will produce, in the months prior to Lent, output in excess of current need, thereby avoiding having to drastically increase output during Lent. A hazard of the balanced output philosophy is that if indications of a decline in demand are ignored and stocks reach an excessive level, the production facility would have to be closed for an extended period. This would severely disrupt the management/employee relationship, which in turn can damage productivity over the long term. Conversely, a failure to increase production when there are signs of improved future demand, may be the cause of 'lost' sales if the production department is unable to keep pace with the need for increased output. Furthermore, if no plans are made to expand the workforce under these circumstances, a point will be reached where existing employee productivity begins to decline, as these individuals reach the point where long periods on compulsory overtime begin to affect morale.

In those sectors of the industry involved in handling a seasonal catch, the objective of stabilised production is impossible. Also many of these sectors now face an additional restriction on the work period because Governments have reduced the catch season as a mechanism to preserve stocks (eg the halibut and herring fishery on the North American west coast; the mackerel fishery in Britain). Production managers then face the unenviable task of hiring and managing a workforce who may only be employed for a 2-3 month period every year. Under these circumstances, the only route to stabilise output may be to introduce techniques to preserve the fish at time of landing by freezing, or storage in brine.

The preserved material can then be used during the rest of the year as the basis for further processing activity. Unfortunately, the solution of using the plant for other types of food processing activity is usually not possible because of either the location of the plant, or the unsuitability of layout for handling non-fish food products. Under these circumstances the cost of an under-utilised fixed asset can only be reflected in higher prices being charged for the seasonal output.

A problem frequently encountered in aquaculture production management is that, because of the biological characteristics of the species being raised, there are one or two brief periods of intense activity at harvest time followed by up to 6-8 months of relative inactivity. This again represents an ineffective use of the production resource, and some companies are researching ways in which they can achieve a year-round output capability. Opportunities include the introduction of additional species, modifications of the culture environment to induce spawning outside of normal seasonal patterns, and the genetic manipulation of brood stocks to alter life cycles. To date these endeavours have met with only limited success, but as more companies move into using intensive or closed systems, a rapid improvement in output smoothing can be expected. Until recently, aquaculture has been restricted in this area by the fact that lengthy periods of experimentation were necessary in order to find effective ways of extending the production season. Some companies now have sufficient background data, however, to be able to use computer models to simulate changes in environmental conditions and thereby isolate, more rapidly, those variables which control biological seasonality. These computer models are also being used to identify the cause of actual output problems which may occur during the growing season, and assist the earlier introduction of actions capable of improving productivity. Industry observers often place emphasis on the capability of aquaculture to produce species at cost lower than that achieved in traditional fisheries. In reality, a more important contribution by this sector may be the capability to provide output at times other than those in the traditional fisheries. This would lead to a more stable supply of fish and seafoods throughout the year which would offer major economies in the area of storage, and permit massive expansion in the sale of fresh products.

Many of the senior management in the industry started their careers in production management, and they are justly proud of the industry's ability to develop innovative solutions to production problems despite the constraints of difficult working conditions, obsolete equipment and limited financial resources. Nevertheless, these individuals frequently fail to recognise the opportunities to introduce new technology and/or monitor scientific advances elsewhere in the food industry which could have applications within their own sector. Effective production management is increasingly dependent upon senior personnel placing greater emphasis on understanding new technologies as the basis for introducing capital intensive equipment into their plants as a path to reducing production costs. Although major innovations have occurred in the last twenty years (eg the introduction of automated filleting machines; computer controlled breading and batter mixing systems), there is still a widespread belief that production management in the industry should concentrate on using their existing equipment and labour force skills as the foundation for increasing productivity. This attitude must change if the industry is to continue to compete on a cost-effective basis with other protein foods. One only has to visit a plant in the soya protein industry or review the literature on single cell protein research to recognise that, unless the fishing industry begins to increase investment in technological innovation, then by the end of this century there are certain production processes which will be unable to compete on cost against other food alternatives. This change must also be accompanied by a less myopic view by companies. The industry is really a subsegment of the food industry, not an autonomous system. An excellent analogy to this situation is the railway industry which in the early part of this century focused its attention on being an efficient operator of rail systems and ignored the realities of being part of a transportation industry. Other areas of this larger system used technology to at first compete for customers and eventually overtake railways as the primary transportation method in most countries.

12
Managing for quality in production

The objective of quality management is to ensure that control systems exist to ensure all the activities of the company result in the customer receiving value for money. The main focus on quality control, however, will be within the production operation because it is here that the majority of actions occur which can affect the physical performance of the product.

The fundamental issue underlying quality management is to deliver what the customer requires; there is little point in establishing a standard that is incompatible with market need. Therefore when designing a new product, the parameters by which the customer will judge the product must be researched. One US shrimp processor assumed, for example, that the only way to meet the requirement for portion control in the food service market was by establishing a quality control specification that peeled and deveined shrimp should be packed using tail count increments of 5. Thus the product range was available in 10-15, 15-20, ... 55-60 count. Above 30 count tails this specification could only be achieved by a double grading production process which increased costs by almost 10%. Subsequent market research to explain poor sales revealed that above 25 count shrimp, serving costs were more important than product uniformity appearance. Hence the solution was to relax the quality control standards and reflect the cost saving in a lower price.

The quality of the finished product in the fishing industry is, to a large extent, a reflection of the quality standards for the raw materials. Thus at the design stage if careful attention is given to defining the correct grade of raw material, many of the quality problems which could occur in the finished product will be avoided. Prior to assigning quality standards, the new product should be put through the processing line and an evaluation made of any possible variations in quality. Raw fish will vary in water content, flesh consistency, smell, taste and colour, and therefore a specification for the final product must be established which accommodates these variations. This is done by creating a tolerance on either side of the defined standard for the product which will be acceptable to the quality controller (eg a breaded oyster may have specification of 50% oyster flesh ± 5%). Once such a standard is set, however, it is vital that the quality control group have sufficient authority to enable them to demand a strict adherence to the specification. The reason for this is that given a choice between achieving output targets and meeting quality standards, production personnel will tend to take short cuts on quality in order to meet production targets. In an ideal situation, quality and production personnel should co-operate with each other because they appreciate that the outcome of a good working relationship is satisfied customers. Unfortunately, however, there are still many companies in the fishing industry whose philosophy appears to be output at any cost. In this situation, the company workforce realise that quality control is a gesture rather than corporate policy and, over time, quality standards decline. It is more than coincidence, nevertheless, that the successful companies in the industry tend to be those who practise a rigid adherence to quality and who are constantly seeking to improve their performance in this area of activity.

Regrettably even in some major companies, quality standards have tended to suffer over the last ten years as organisations have struggled to hold costs down in the face of rising raw material prices. A common solution has been to move to a lower grade of raw material or to increase the proportion of low cost ingredients in the product specification (eg move from whole to minced fish in fish fingers; increase the breading level in breaded scampi). These actions have sustained profit margins but in many instances the consumers have identified the deterioration in quality and gradually reduced their per capita consumption level. Certainly when one examines the causes for the decline in unit sales volume for certain coated fish products in North America over recent years, there is little evidence to suggest that rising prices was the real reason for the fall-off in sales.

PRODUCT INSPECTION

The inspection of materials should begin at the 'goods inbound' stage before they are used for further processing. In the fishing industry, the decision on whether or not the material is satisfactory is frequently based on visual examination. For example, a company purchasing bulk pack, peeled shrimp for repacking into retail pack sizes may open a few cases, and if the shrimp appears to be the right size and colour, the entire shipment is accepted. The drawback to this method is that the inspector has no defined standards against which to rate the material, and product of marginal quality might be accepted which could lead to problems later in the processing operation. Also if product is rejected and the inspector simply returns the shipment as 'judged unacceptable' the supplier is given no lead as to what specific improvements are required in order to avoid similar problems in the future.

To overcome the weaknesses of visual inspection, some companies have developed ranking systems for important variables (eg frozen fish blocks rated for flesh consistency, water content, flesh colour, smell, uniformity and absence of harmful bones). This approach known as 'measured-variable inspection' tends to be extremely effective and, once understood by suppliers, can often lead to a reduction in the level of materials rejected at the goods inward stage. Development of appropriate measured-variable schemes may be based upon parameters established by a single company, but in some countries attempts are being made by Government inspection services to define minimum standards for key variables in the hope that these will be adopted on an industry-wide basis. Even with measured-variable inspection, many of the results are based upon semi-qualitative judgments by the quality inspector, and although attempts have been made to introduce scientific, machine-measured evaluation systems, to date the natural variability of fish flesh has been a major obstacle in their development.

The increasing international trade in the fishing industry has created problems for importers who are forced to reject materials at the port of entry. It would clearly be more efficient for any rejection to occur before the product leaves the country of origin. In recent years, therefore, some companies have begun to work with their suppliers to strengthen the latter's quality control procedures, and this has often reduced rejection levels to almost zero, thereby benefiting both parties.

The next part of an inspection procedure is on 'work in process' materials, where product is removed on a regular basis at each step in the production process for quality evaluation. The purpose of this activity is to pick up a problem in processing at the earliest possible moment, before large quantities of faulty goods have been produced. For example, on a breaded fish portion line, the unbreaded fish would be evaluated for weight and size, further samples taken after breading and frying to assess the level of coating and appearance of the finished product, and finally a sample taken after freezing to assess the product before it is packaged. If any quality defects are identified, the production personnel can move to rectify the problems before large quantities have been produced. Work in process inspection should be constantly carried out. In this way the quality control group have continuous records of performance which can be used to identify trends in performance that can be indications of a developing problem before product quality becomes unacceptable. These records are also vital if a customer complaint occurs. For if the company is certain that a complaint is not due to a production fault, this will greatly reduce the task of identifying the true cause of the problem. It is not unusual to discover that the complaint stems from an error made after the product left the factory due to incorrect handling procedures by intermediaries or the final customer.

As a final safeguard against defective goods reaching the customer, the quality control group should carry out a regular programme of post-production inspection. This should not just be done on product as it leaves the processing area, but also on finished goods which have been in inventory for some time. Given the highly perishable nature of fish and seafoods, it is frequently the case that defects are caused more by incorrect handling during storage or distribution, than on the production line. Furthermore, a company's responsibility for quality management does not cease once the product has left the plant, but remains up until the point of final consumption. It is therefore advisable to arrange for samples to be taken at various points in the distribution system to ensure that intermediaries are handling the product correctly. Where problems do become apparent (eg a wholesaler who is not storing product at a sufficiently low temperature), the quality control personnel may consider mounting seminars for members of the trade to improve their knowledge of how to care for product and avoid quality deterioration.

Every company in the fishing industry must be aware that they have a responsibility to ensure that product contamination, which could cause illness or even fatalities, does not occur. It is the quality control group's responsibility to ensure there is a high level of personal hygiene among employees. This is achieved by monitoring employees' adherence to regulations over washing before handling product and ensuring that appropriate protective clothing is worn at all times. Where the plant has special crews for cleaning processing areas during or at the end of a production shift, their activities should be executed under the direction of quality control personnel. These individuals will also be responsible for monitoring bacteria levels by taking and culturing samples from the line and inspecting the plant to identify any shortcomings in the cleaners' activities.

An event most feared in the fishing industry is an outbreak of food poisoning. Public health officials can only contain such an incident if they receive full co-operation from the company whose product is the cause of the problem. It is the responsibility of the quality control group, therefore, to establish and maintain a record system which describes to where and when all products are shipped. This task is made much easier if the company has installed a product coding system which can be used to identify date of manufacture, quantity produced and final destinations. In this way, should a quality problem occur and there is a possibility of a product recall, the company can rapidly identify the quantity involved and the intermediaries to contact. In some countries there is legislation on product coding and recall procedures, and companies are legally obligated to co-operate with public health officials. Where these systems do not exist, it is advisable that companies, through a trade association, attempt to get all member organisations to participate in a voluntary programme of quality standards and product coding. The natural inclination of senior management is to avoid any activity which will increase costs, but it must be recognised in those areas where legislation does not exist, that when a Government is forced to intervene, the costs of complying with mandatory regulations are usually much higher than for a voluntary intra-industry scheme.

QUANTITATIVE ANALYSIS

Record keeping and analysis are the keystones to an effective quality control system. The analysis techniques must accommodate the natural variations in raw materials by the establishment of control limits outside of which products will be rejected. This situation, and the fact that, in most instances, it is only feasible to inspect a small proportion of total production, means that quantitative quality control methods rely heavily on the use of statistical procedures to assist in reaching an acceptance or rejection decision.

One approach is to set a maximum level of defective items that is acceptable within a sample. If this level is exceeded then the whole production batch is condemned. This is known as 'acceptance sampling by attributes', and the acceptance level is calculated using a binomial distribution curve which assumes a statistical relationship between the percentage level of defective items in a batch and the probability that the batch will be accepted. Although the method is widely used, the maximum acceptable level of defects is an arbitrary judgment and therefore does not adequately allow for the occurrence of natural (or 'random') variations. To allow for variations it is necessary to monitor production on a regular basis and compare the results against predetermined upper and lower control limits. Results outside these limits will indicate the need to take immediate action. As most quality problems develop on a gradual basis, however, the establishment of upper and lower warning points within the control limits means that if results indicate a quality variation between the warning and control limit, remedial action can be taken before it is necessary to initiate complete rejection.

This type of control system is usually based on the assumption that results from samples will fit a normal probability distribution where 99.8% of change or random variation can be expected to fall within 3.09 standard deviations of the mean value for the sample. Thus any sample value falling outside the boundary of 3.09 standard deviations from the mean can almost certainly be assumed to indicate an actual quality problem. Upper and lower control limits set to exclude only 0.2% of observations being caused by random variation will be stated in statistical terms as the control limit being equal to $\overline{X} \pm 3.09$ standard deviations where $\overline{X}$ is the mean value of the sampled variable. The warning limit can then be set to exclude 5% of observations being caused by random variation by setting this limit equal to $\overline{X} \pm 1.96$ standard deviations.

THE COST OF QUALITY

The management dilemma is that it is very easy to measure the cost associated with operating a quality control system but very difficult to assess the financial benefits. The costs include the prevention procedures (eg the amount of ice used by a trawler), the running expenses for a laboratory, and the revenue from rejected or re-worked product. A possible objective for companies positioned in the quality segment of the market is that of 'zero defects' on the grounds that any lower standard would endanger the company image. Even in these companies, the quality control personnel must recognise that they are as much responsible for generating profits as any other department. This can only be achieved if they execute a continuous assessment of their standards and analyse the costs and benefits of their programme. In this type of assessment, the issues to be addressed should include:

- How does a standard contribute to the value of the product?
- How much does a standard add to the production cost?
- Can a lowering of a standard be introduced without significantly reducing quality?
- Can a specific quality problem be overcome by a change in the raw material specification or the processing technique?

The first step in such an evaluation is to collect information costs, customer requirements and possible future changes in production technology. These data can be reviewed with other interested departments, usually production and marketing, and after discussions, alternative product specifications can be prepared for review with management.

Within the fishing industry the role of quality control is still perceived to be a source of increased costs, rather than a centre of knowledge which can be utilised to initiate changes capable of improving profitability. However, for this very radical philosophy to gain acceptance, senior management in the industry will have to more readily accept that quality control is a mandatory component of effective production management. In turn, quality control personnel must acquire greater understanding of the wider contribution they could be making to the overall company operations through a more enlightened approach to possible contributions in the area of strategic planning. A tragic example of failure in this area often occurs in the aquaculture industry. An important feature of many culture systems is that the quality of output at harvest is significantly higher than for the same species landed from a traditional fishery. The reason for this is that the catching operation stresses the catch by compression of materials in the net or by direct damage from hooks or gaffs. Yet many of the handling procedures recommended by quality control personnel in aquaculture are only designed to sustain a product quality standard similar to that achieved in the traditional fishery. For example some shrimp farms hold harvest on ice for up to a week in order to accumulate sufficient shrimp for a full truck load shipment for the processing plant. The holding period is justified on the grounds that even after a week, the quality of the shrimp will still be equal to that landed by a trawler. In reality, these farms are losing a major opportunity to strengthen their market position and receive a premium price by producing a far superior product to that obtainable by trawling.

13
Managing human resources

RESOURCE PLANNING

Along with the developing philosophy of strategic management has come the recognition that the company's personnel are probably its most important resource. This has led to the expansion of the personnel management role from that of the hiring, paying and firing of employees to that of the responsibility for the optimal current and future use of the human resources within the organisation.

The first stage in the human resource management process is to translate the company strategic plan into a definition of the employee skills required to execute the plan within each of the functional areas of the business. The complexity of this task has been significantly increased because skill definitions have to reflect the expected influence of changing technology on the company. In the production department, for example, in the fifties and sixties, many fish plants used a large number of unskilled manual workers with a small number of supervisors, most of whom had acquired the knowledge and supervisory skills from working on the production line prior to their promotion. The interior of a modern processing plant is dramatically different, with many tasks now automated (eg filleting machines; microprocessor controlled instrumentation for automated machine adjustments; computerised mixing and bulk material handling systems). These changes have reduced the need for unskilled labour, but have increased the need to employ supervisors with formal qualifications in engineering and electronics. The same situation is also true in the catching sector of the industry with the new large trawlers requiring the skipper and crew to work with sophisticated electronic technology to maximise catch per unit of fishing effort.

It is even more difficult to forecast the future skills needed in aquaculture because of the rapid change taking place in the development of improved culture techniques in response to the ever increasing knowledge about the best way to manage species. The speed of this change is illustrated by the move from extensive to intensive grow-out systems in shrimp culture over the last ten years, and in the last five years the use of hatcheries rather than natural stocks to supply adult brood stocks. These changes are reflected by over a 50% reduction in the need for unskilled labour for pond management and in an average sized facility, an increase of over 1 000% in the technical staff needed to manage the more sophisticated culture techniques. This trend will continue as more operations move into closed system shrimp culture.

The skills required have to be matched against the company's current personnel inventory to identify areas of imbalance requiring immediate or future attention. To sustain an effective management/employee relationship, it is preferable to remove these imbalances by retraining current staff where possible, rather than replacing them with new employees. Where analysis indicates a need for additional or different staff in the future, the personnel department is responsible for forecasting the availability of such individuals in the labour pool market. In the past, the fishing industry has not generally been affected by an inadequate supply of labour because plants and catching operations were usually located in areas of the world where few alternative employment opportunities existed. This is no longer the case in many areas, however, because the supply of labour has been influenced by the following factors:

- Workers have been attracted by other industries competing in the same labour market (eg the oil industry in the North Sea).
- Improved communication and transportation has caused a migration of workers from an area (eg from previously isolated Newfoundland fishing out-stations, after the construction of a road).
- In some countries, the earnings at a fish plant are less attractive than Government welfare payments for the unemployed.
- Other industries offer higher earnings and better promotion prospects for skilled workers (eg service technicians for navigational equipment on trawlers leaving to enter the computer industry).
- The uncertainty of employment because hours worked depends upon catch, whereas other industries have a more

stable environment although may pay lower wage rates (eg workers in New Hampshire, USA who leave semi-skilled jobs in fish plants and accept unskilled jobs in supermarkets).

Having determined the relationship between supply and demand for labour, the personnel department can advise management on a programme to resolve any imbalances, and on the anticipated costs for executing company strategies. It is not unknown for a resultant wage forecast to reveal a future operating costs problem, and in this situation the personnel group in conjunction with the affected departments, will have to examine alternative employee strategies. Only a minority of companies in the fishing industry produce long-term wage and salary forecasts, but as this is often the largest single cost component in an organisation, this lack of planning must be judged very shortsighted. If a company waits until costs exceed revenue, it is usually too late to introduce an alternative plan. A notable exception to this generalisation about the industry is to be found in Japan where companies have paid careful attention to future wage costs, introduced revised processing technology where appropriate, and in certain instances relocated an entire sector of operations to another country when labour costs could not be stabilised by automation. The western firms have a much poorer performance record in this area, and in some countries the lack of recognition of upward trends in operating costs has eventually led to plant closures and bankruptcies.

HUMAN RESOURCE ACQUISITION

It is often stated that recruitment is the most important aspect of the personnel function because inefficient hiring results in poor employee performance, defective products, absenteeism, dispute, and ultimately a high labour turnover. Hence the objective in recruitment is to match the defined role in a job, both current and future, against the available applicants. Where an exact match is not attainable, both the job specification and the procedures for attracting applicants must be reviewed in order to resolve the incompatibilities, because compromise is not acceptable.

Ultimately the selection process is dependent upon the judgment of one human by another. However, if a systemised approach is used, it is possible to minimise the risk of error. The first step in a systemised approach is to collect information on the nature of the vacancy to be filled. This analysis will lead to an accurate job description which can be illustrated by that for a laboratory technician in a fish processing factory:

Title	Laboratory technician
Department	Quality control
Reports to	Head technician
Duties	Take samples from processing line Analyse samples relative to product specifications Advise head technician of any variance in samples Keep laboratory analysis records Analyse trends in processing performance Maintain laboratory supplies and recommend re-orders
Occasional tasks	Liaise with Government inspection service Conduct plant tours Attend production liaison meetings Run classes on hygiene regulations for plant employees
Salary	Monthly
Hours	8.30-12.00: 1.00-5.00; 5 days per week Additional hours when the plant is working overtime
Possible change in job role	Greater emphasis on automated analysis techniques

From the job description, it is a logical step to define an outline definition of the type of person best suited for the post. This specification can also be used for performance evaluation, promotion and training after appointment of an applicant.

Most jobs require personal skills to relate to, and work with, other employees. This aspect should also be included in the personnel specification and will form part of the basis for reviewing applications and selection at interview. For the laboratory technician, the personnel specification would cover the following areas:

Job title	Laboratory technician
Department	Quality control
Physical	
Age	21-45 (desirable)
Sex	No preference
Height	Unimportant
Health	Must be physically mobile, not suffer from any communicable disease
Attainments	
Examinations	Passes in school leaving examinations at least in English, Mathematics and Science
Experience	Minimum of two years as a technician (essential) Knowledge of food industry analysis methods (preferred)
General intelligence	Should show at interview that the person is realistic and knowledgeable about the technical aspects of quality control work
Special skills	Evidence of social skills as indicated by (a) understanding of relationship obstacles of quality control and production personnel and (b) general appearance and attitude

In preparing a personnel specification, it must be recognised that in many countries Government legislation exists on hiring practices, discrimination against minorities, and contractual obligations of employers to employees. Without attention to these issues the company may risk being involved in expensive legal penalties for incorrect personnel practices.

The other essential factor of the recruitment process is the method by which qualified applicants are attracted. This is done by various means including advertising, employment agencies, notification to existing employees, and in the case of senior managers, possibly through the use of personnel consultants. Having drawn up a short list of applicants, the final selection decision is made by interview. The form of the interview process is extremely varied. The ultimate test of whether the company's interview practice is effective is whether the applicants are successfully matched against the job specification, and if the new employee's performance is satisfactory. In the event that labour turnover is high or productivity low, this would indicate a recruitment procedure problem and it may be necessary to obtain specialist advice to develop a solution.

HUMAN RESOURCE DEVELOPMENT

The quality of employee performance is dependent upon the company's efforts to fully exploit their existing skills and to provide opportunity for the acquisition of new skills. This can only be achieved by the establishment of structured training programmes across all departments and at all levels within the company.

The minimum requirement for a training programme is an induction course to integrate new employees into the company as rapidly as possible. Issues covered should include a review of the company history, plans, products and facilities; a definition of the employee's responsibilities; and a description of pay structures, rules of conduct, arbitration procedures, benefit programme, safety regulations and opportunities for advancement. This form of orientation should occur at whatever level the employee is joining the company, be it as a cleaner or line manager. Obviously, however, the content of the orientation programme should be designed to reflect the capabilities and motivation of the participants. Whether or not job skill training is necessary at the point of induction will depend upon the existing skills of the employee in relation to the job specification. A widespread habit in the fishing industry is to place new operatives (both in the administrative departments and plant workers) next to an experienced employee to observe how the job is done. The two errors in this approach are (i) the assumption that the experienced operative is doing the job correctly relative to company procedures and (ii) the new employee, in the desire to please, will often not be willing to question his colleague about all the aspects of the job.

The more enlightened organisations have invested in the design of an instruction manual of job skills, or a specifically nominated individual - often the first line supervisor - to take the new employee through a structured learning programme. Performance is measured at various stages in the instruction period using standards based upon experienced workers' average performance levels. Where shortcomings are identified, additional training may be given. There is strong evidence that where this type of induction programme has been installed, new operators reach optimum efficiency in a shorter period and, more importantly, fewer workers leave soon after joining the company because of discomfort with their new job role. In certain situations, it is also necessary to provide guidance on personal hygiene in order to avoid product contamination. This tends to be more prevalent in underdeveloped countries where employees have been socially deprived and have no effective sanitary facilities in their homes.

Those individuals who join the company at or above the first line supervisor level with previous work experience are usually thought in the fishing industry to require no further training beyond an orientation programme on company policies. This reflects a prevailing general opinion that managers at all levels are fully qualified and require no further training. Given that a company is only as good as the management, this lack of recognition of the need for on going systemised training to improve all employees' skills, knowledge, insights and attitudes, is doubtless an important cause of poor management performance in the fishing industry.

To be an effective manager, individuals need skills in the areas of technical proficiences specific to a business function, inter-personal skills to work effectively with or to direct others, and conceptual skills to perceive the organisation as a whole, recognise the interrelationships of variables, and guide their area of responsibility along a path of success. As a manager moves up in a company, his technical skills become less important, whilst effective conceptual skills become more so. The requirement to manage others is present at any level. It is extremely rare that an individual can develop this skill without being involved in a structured development programme, and it is the responsibility of the personnel department to ensure that this exists.

Senior management must be involved in the setting up of any management development programme because their input is required on the direction executive development should take. Furthermore, their overt support is needed to ensure that the programme receives complete acceptance at every level of management. All too often the attitude of senior managers is that they reached their position in the company without these activities, therefore they are not necessary for the lower level executives, especially as the latter have attended a college before joining the organisation. In these circumstances, it is necessary for the personnel director to explain to the board the need for further training to cope with the complexity of the modern business environment and the impact of new technology on employee performance. The director demonstrates the value of training programmes by presenting the evidence from other industries that such activities are often the source of the most innovative revisions in corporate strategies. It must be realised by the Board, however, that the benefits of investing in further training for managers will not be apparent immediately, and that it may take 2-5 years before quantifiable benefits can be identified.

The start of the management development programme will be an appraisal of the organisation's current and future management capability as it relates to the existing skills of the management team. It is probable that weaknesses of ability will be found in areas such as modern theories of management, human resource utilisation, personal communication and business planning. It is rare that managers have technical weaknesses in their area of functional specialisation, although this may occur where the department is heavily affected by a rapid technological change (eg the influence of computers on the accountancy function). The responsibility for the nature and type of training will depend upon the internal resources of the personnel department. In larger companies the personnel group may contain staff with experience in many of the training issues to be covered. Within the fishing industry, given that on average both companies and their personnel department are both small, the most practical route would be to send managers on courses or to bring in an advisor to instruct certain managers to perform the training function within the company.

Management development programmes are expensive, and one possible solution for meeting the cost is for the training to be organised by the trade association, with members contributing to the costs in direct proportion to the number of their executives who participate. Another approach is to involve Government training agencies or international organisations such as the FAO. Any investment represents a deferment of immediate profit to gain incremental return from assets in the future. Investing in the improvement of executive performance is still an unfamiliar concept in the industry. Unless it begins to gather a more widespread acceptance, then human resources like other industry assets will depreciate in value and over time cease to generate an adequate return.

14
Industrial relations

WAGE AND SALARY ADMINISTRATION

From an employer's point of view, the payment of wages is a business cost, and the objective of cost minimisation will include, therefore, keeping labour rates as low as possible. Employees on the other hand, view wages as the means of meeting their financial needs in society. They also compare their income to others both inside and outside their place of employment to assess the fairness of the wage scale for the work they do. This difference in perception between employer and employee is often the basis of industrial disputes. In addition, as communication and education has improved, the workforce in most countries consider now that negotiations over wages should replace the more traditional practice of employers setting wage scales without any discussions with employees. The fishing industry has tended to resist this social change, and during the last twenty years there have been severe dislocations in productivity as a result of prolonged, and often bitter, disputes over wages. If the industry is to be less affected by the problem in the future, management must recognise the need for a more modern, comprehensive approach to wage and salary administration.

The goals for any compensation programme must be to:

(1) Recognise that rates of pay serve to allocate people within the labour market by defining the perceived attractiveness of various jobs. The firm must therefore be competitive in their wage rates if it wishes to recruit employees.
(2) Create a job evaluation, pay scale and employee classification procedure which enables management to control costs and prevents supervisors authorising increases in size or frequency which could circumvent costs control.
(3) Ensure that the compensation programme is perceived by

employees as fair and equitable across all departments, thereby minimising grievances, complaints or resignations.

(4) Act as a motivator to increase productivity by acting as an incentive. (Within the context of this fourth objective it must be recognised that wages and salaries - although important - are only one part of the employee motivation process. Once basic financial needs are met, other factors such as security, power, achievement, social position, etc also come into play and represent other mechanisms by which employers can stimulate employee output.)

A widespread lack of understanding in the fishing industry of the breadth of the compensation programme objectives is underlined by comments at trade association meetings such as 'we pay the lowest wages possible' or 'just enough to keep the workforce and the unions out'. It is therefore the role of the personnel directors in the industry to introduce senior management to a more logical, systematic and equitable approach to compensation if the trend of increasing wage disputes in many parts of the world is to be reversed.

The firm has to first ascertain the prevailing rate for the same class of work in other firms and for the other industries competing in the same labour market. In the event that this investigation indicates an inability of the company to compete, the short term solution may be to pay a lower rate, but accept that this will usually be accompanied by poor productivity. Over the long term, management must examine alternative production technologies or consider plant relocation if the company is ever to perform at a high level of profitability.

Within the organisation a clearly understandable policy must be developed on the nature of every job, including a statement of the personal qualities required, accompanied by a system whereby the skill content for all jobs can be measured. There are job evaluation methods which either use qualitative lists against which each job is assessed or numeric scales which permit weighing of the various skill elements on a quantifiable basis. Whichever technique is used, the ultimate purpose is to determine the skill content of each job which can then be used to calculate a range of wage rates within the company. These wage rates should also contain some mechanism to permit increments within each pay scale for factors such as length of service and merit. When pay is related to merit, the company will also require a formal programme for appraising employee performance on a regular basis. This should be executed through a performance review meeting with all employees explaining why an individual's performance qualifies or disqualifies them for a wage increase. It is more difficult to apply this type of review to supervisors and managers because their role may be less easy to assess on a quantifiable basis. One solution to this problem which has gained popularity over recent years is a concept known as 'management by objectives' (MBO). With this system, the individual and his immediate supervisor determine objectives of performance for a defined period, and at the end of this time, the employee's attainments relative to these targets are reviewed.

Management always finds appeal in a wage scheme which links earnings directly to output. To be effective a wage incentive of this nature requires that:

(1) The output of employees must be measurable and suitable for standardisation.
(2) There must be a consistent relationship between the level of skill and amount of effort exerted by employees and output.
(3) Work must be readily accounted to the proper individuals, or if this is not feasible, some mechanism must exist to use a group, department or plant-wide incentive scheme.
(4) There must be an opportunity for reducing the unit cost of the product or increase productivity per worker. This is mandatory because the only logic for installing the system is that it offers management a way to increase profits at a rate equal to or greater than the incremental costs of the incentive payments.
(5) The system must be acceptable to the workforce.

The requirements for such a scheme demonstrate the need for extreme care over the decision to introduce an incentive scheme and their popularity amongst both employers and employees seems to vary over time and between countries. Although there is little published information on schemes in the fishing industry, it would appear that they are popular in the primary processing sector in North America, but less common in Europe and in most underdeveloped countries. In these latter areas of the world, it is possible that the average level of education amongst plant workers is insufficient to understand the concept of productivity measurements. This will mean that no benefit will accrue from using a wage scheme linked to this type of incentive.

HEALTH AND SAFETY

Employer attitudes about safeguarding the health and safety of staff have improved beyond all recognition since the days of the industrial revolution. Nevertheless in terms of costs, many managers still tend to only evaluate the direct costs of paying employee benefits of the premiums for health and safety insurance. In fact the indirect costs such as lost output, time spent by supervisors resolving problems, damage to equipment, and the administrative expense have been estimated to be four times higher than the direct costs. Hence as a component of the objective to minimise labour costs, the personnel department must ensure that employees' welfare at work receives strict and detailed attention.

The fishing industry is indeed fortunate in that most of its processing technologies do not represent unusually high risk in terms of occupational diseases. Hazards of this type are classifiable into physical (eg noise, heat vibration and radiation), chemical (eg dusts, poisonous fumes, gases, toxic metals and carcinogens), biological (eg bacteria, fungi and insects) and stress (eg physical, psychological and ergonomic). Many of these hazards can be avoided by adhering to Government or international commissions' recommendations on general safety procedures which apply in the food industry. Nevertheless managers in the fishing industry should understand that change is often the cause of new occupational disease risks (eg increased noise levels when moving to automated equipment; introduction of new chemicals for disease prevention or treatment in aquaculture). Hence careful attention should be given to these issues when new schemes are introduced and, furthermore, a nominated individual should be responsible to ensure employee adherence to disease prevention policies.

In the area of industrial safety the fishing industry faces the burden of lives lost at sea due often to sudden changes in weather conditions. Nevertheless, the industry in certain parts of the world could reduce accident levels if a more positive attitude to accident prevention was adopted. Where companies have realised that reduced accident levels will reduce costs, significant advances have been made in reversing trends in accident rates. The fundamental step is to 'design out' any potential hazard in equipment or processes. Where this is not practical, then risk areas should be protected by guard rails or shields and employees issued with protective clothing. In addition, to achieve safety consciousness, employees should be involved in a regular programme of safety education. Finally, it should never be assumed that no further improvements in a safety programme can be made. All accidents should be investigated to isolate the cause, and if necessary revisions made to avoid repetition in the future.

GRIEVANCE PROCEDURES

Even in the best-managed company, situations arise when employees believe that they have been treated unfairly. This perception of injustice, if there is no channel through which it can be aired, will eventually lead to frustration and a decline in productivity. In the fishing industry, the commonest solution is to leave handling of the problem with an employee's first line supervisor. Evidence from other industries, however, indicates an overwhelming trend to introduce more formal procedures, because complaints can often not be settled by the supervisor. Possible reasons for this may be the supervisor's lack of formal training in personnel management, his probable lack of real authority to instigate the action necessary, and the attitude of the employee towards the supervisor who is frequently the original cause of the grievance. Furthermore, as senior management are not included in the procedure, they are unaware of employees' problems and therefore unable to operate their departments at maximum efficiency.

In a dispute between an employee and the immediate supervisor, neither party can be considered impartial, and hence any formalised grievance procedure must seek to introduce an element of neutrality. Where the company is unionised, impartiality is relatively simple to achieve because the union, as part of the collective bargaining agreement with management, will have access to arbitration from outside the company. In these situations, the grievance procedure is standardised into a series of defined stages. The first step is usually for the employee and a union representative to meet with the supervisor. If this does not resolve the problem, the dispute will be moved up the system by involving a representative of the personnel department, then the personnel director, and eventually can involve a member of senior management. If the management's decision is not acceptable, the final step would be for the employee to request that the grievance be taken to outside arbitration. For this system to work effectively, however, all parties must accept that the arbitrator's decision is binding.

In non-unionised fish companies it is rare to find a formal grievance procedure. In general, management have chosen to ignore this important aspect of labour relations, although many would claim that they operate an 'open door' policy through which employees can air their complaints to senior management. This rarely works because of the social and organisational distance between most workers and the head of an organisation. Furthermore if a senior manager overrules a supervisor in these circumstances, the latter's day-to-day authority would be undermined. Where possible, non-unionised companies should establish a structured grievance procedure including access to arbitration, possibly through the services of the trade association or a Government agency.

The ultimate objective in personnel management is to create a system of policies and procedures which are understood by all employees, thereby avoiding any situation which causes individuals to decide whether they have been unfairly treated. In establishing guidelines, it is vital to involve representatives from all levels of the company, and the system must be carefully monitored to ensure consistency in the treatment of offences. Firing is the ultimate punishment, which is sometimes known as 'industrial capital punishment'. No company should be proud of a high record of firings. Indeed, rigorous employee procedures coupled with an effective labour relations programme should almost eliminate the need for any employee to be discharged.

UNIONS

The desire for self expression is a fundamental human drive, and employees need to know that their employers are willing to listen to them. By forming a union, employees can create a mechanism through which the consensus of their opinions can be communicated to management. Typically there is a local union in a specific factory or city, which, along with other 'locals', is affiliated to a national union which represents the membership on a national basis.

In the fishing industry the level of union membership varies by country, being probably, in the free world, highest in Europe and lowest in underdeveloped nations. Management's attitude in general is that unions are undesirable as they are simply out to get all they can for their members. This is somewhat naive because of course the union's role must be to strive to improve conditions of employment. Principally this is achieved through collective bargaining where the union seeks to reach agreement with the employers by negotiation on wage rates, hours of work, and any other issues which affect their members. The final agreement may apply to a single plant or, in certain countries, to all companies in the industry.

Once a union has won bargaining rights for the employees, management will often find that the way they handle industrial relations is considerably altered. No longer are they able to make unilateral decisions on wages, conditions of employment and employee discipline. This will restrict management's freedom of action, and this reduction in flexibility over moving employees from one job to another or revising manning levels can impair company efficiency. Obviously the average manager given the choice between autonomy and industrial democracy will tend to favour the former. On the other hand, it is fair to say that the presence of a union encourages management to become conscious of the employees' viewpoints. Any omission in a company's personnel policies will quickly be identified by the union and remedial action instigated. Thus the advent of a union has caused some companies in the fish industry to introduce supervisor and executive training programmes focusing on human resource management, with the aim of improving labour relations and reducing union influence over their employees. This action is in fact beneficial to both employee and employer because the resultant improvement in the work environment is reflected by higher productivity. If more companies understood the benefits of good industrial relations, 'there would' to quote one union official 'be no point in having a union'.

In the past there have been strong attempts by companies in the industry to collectively oppose unionisation. The union's response was usually to bring one key plant out on strike as an example. Both sides would subsidise the affected plant, the company receiving financial aid from other companies, the employees being supported from union funds. It is certain that whatever the outcome, employees or employer capitulating, no side can be considered a 'winner' in this situation. Once both sides have acknowledged this fact, there has been an increasing tendency for unions and management to realise they have a common goal. By working together to identify ways to increase productivity or reduce costs, the company benefits from improved profitability and the employees enjoy greater security plus a wage increase funded by improved performance, not through a reduction in profits.

For this more democratic relationship to exist, employers in the industry must be well versed in modern human resource management practices, and the unions must fully understand the financial implications of their objective to represent the best interests of their members. Regrettably there is little evidence yet to suggest that these philosophies are being adopted by more than an enlightened few in the fishing industry. Until that time, the much needed improvement in the industry's standards of labour management relations seems unlikely to occur.

Other books published by
Fishing News Books Ltd

Free catalogue available on request

Advances in aquaculture
Advances in fish science and technology
Aquaculture practices in Taiwan
Atlantic salmon: its future
Better angling with simple science
British freshwater fishes
Commercial fishing methods
Control of fish quality
Culture of bivalve molluscs
Echo sounding and sonar for fishing
The edible crab and its fishery in British waters
Eel capture, culture, processing and marketing
Eel culture
Engineering, economics and fisheries management
European inland water fish: a multilingual catalogue
FAO catalogue of fishing gear designs
FAO catalogue of small scale fishing gear
FAO investigates ferro-cement fishing craft
Fibre ropes for fishing gear
Fish and shellfish farming in coastal waters
Fish catching methods of the world
Fisheries of Australia
Fisheries oceanography and ecology
Fisheries sonar
Fishermen's handbook
Fishing boats and their equipment
Fishing boats of the world 1
Fishing boats of the world 2
Fishing boats of the world 3
The fishing cadet's handbook
Fishing ports and markets
Fishing with electricity

Fishing with light
Freezing and irradiation of fish
Freshwater fisheries management
Glossary of UK fishing gear terms
Handbook of trout and salmon diseases
Handy medical guide for seafarers
How to make and set nets
Introduction to fishery by-products
The lemon sole
A living from lobsters
Making and managing a trout lake
Marine fisheries ecosystem
Marine pollution and sea life
Marketing in fisheries and aquaculture
Mending of fishing nets
Modern deep sea trawling gear
Modern fishing gear of the world 1
Modern fishing gear of the world 2
Modern fishing gear of the world 3
More Scottish fishing craft and their work
Multilingual dictionary of fish and fish products
Navigation primer for fishermen
Netting materials for fishing gear
Pair trawling and pair seining
Pelagic and semi-pelagic trawling gear
Penaeid shrimps - their biology and management
Planning of aquaculture development
Power transmission and automation for ships and submersibles
Refrigeration on fishing vessels
Salmon and trout farming in Norway
Salmon fisheries of Scotland
Scallop and queen fisheries in the British Isles
Scallops and the diver fishermen
Seine fishing
Squid jigging from small boats
Stability and trim of fishing vessels
The stern trawler
Study of the sea
Textbook of fish culture
Training fishermen at sea
Trends in fish utilization
Trout farming manual
Tuna distribution and migration
Tuna fishing with pole and line